VIII Brazilian Meeting on Simulational Physics 2015

Journal of Physics: Conference Series Volume 686

Florianopolis, Santa Catarina, Brazil
3 – 8 August 2015

Editors:

N.S. Branco
W. Figueiredo
J.A. Plascak
M. Santos

ISBN: 978-1-5108-2097-5
ISSN: 1742-6588

Printed from e-media with permission by:

Curran Associates, Inc.
57 Morehouse Lane
Red Hook, NY 12571

Some format issues inherent in the e-media version may also appear in this print version.

Printed by Curran Associates, Inc. (2016)

For permission requests, please contact the Institute of Physics at the address below.

Institute of Physics
Dirac House, Temple Back
Bristol BS1 6BE UK

Phone: 44 1 17 929 7481
Fax: 44 1 17 920 0979

techtracking@iop.org

Additional copies of this publication are available from:

Curran Associates, Inc.
57 Morehouse Lane
Red Hook, NY 12571 USA
Phone: 845-758-0400
Fax: 845-758-2634
Email: curran@proceedings.com
Web: www.proceedings.com

Table of contents

Volume 686

VIII Brazilian Meeting on Simulational Physics
3–8 August 2015, Florianópolis, Santa Catarina, Brazil

Accepted papers received: 6 January 2016
Published online: 10 February 2016

Papers

This special issue includes invited and selected articles of the *VIII Brazilian Meeting on Simulational Physics* (BMSP), held in Florianópolis, Santa Catarina, Brazil, from 3rd to 8th August, 2015. This is the eighth such meeting, and the second one to have contributed papers published in *Journal of Physics: Conference Series* (the other was the VII BMSP).

The previous meetings in the BMSP series took place in the mountains of Minas Gerais, in the region of the Brazilian Pantanal, and in the shores of Paraíba. Now, for the first time, the Meeting was held in Florianópolis, with its pleasing shores, the capital of Santa Catarina state.

The VIII BMSP brought together about 50 researchers from all over the world for a vibrant and productive conference. As in the previous meetings, the talks and posters highlighted recent advances in applications, algorithms, and implementation of computer simulation methods for the study of condensed matter, materials, and out of equilibrium, quantum and biologically motivated systems. We are sure that this meeting series will continue to be an important occasion for people working in simulational physics to exchange ideas and discuss the state of the art of this always expanding field.

We are very glad to put together this special issue, and are most appreciative of the efforts of the editors of the *Journal of Physics: Conference Series* for making this publication possible. We are grateful for the outstanding work of the Florianópolis team, for the financial support of the Brazilian agencies CAPES and CNPq, and of the Federal Universities UFPB and UFSC. At last, but not least, we would like to acknowledge all of the authors for their written submissions.

N. S. Branco

W. Figueiredo

J. A. Plascak

M. Santos

Guest Editors

Journal of Physics: Conference Series **686** (2016) 011001 doi:10.1088/1742-6596/686/1/011001

Journal of Physics: Conference Series **686** (2016) 011002 doi:10.1088/1742-6596/686/1/011002

Peer review statement

All papers published in this volume of *Journal of Physics: Conference Series* have been peer reviewed through processes administered by the proceedings Editors. Reviews were conducted by expert referees to the professional and scientific standards expected of a proceedings journal published by IOP Publishing.

Journal of Physics: Conference Series **686** (2016) 012001 doi:10.1088/1742-6596/686/1/012001

Folding in a semi-flexible lattice model for Crambin

Guangjie Shi[1], Alfred C.K. Farris[1], Thomas Wüst[2] and David P. Landau[1]

[1]Center for Simulational Physics, The University of Georgia, Athens, GA 30602, USA
[2]ID Scientific IT Services, ETH Zürich, 8092 Zürich, Switzerland

E-mail: sgjerry@physast.uga.edu

Abstract.
Using the Replica-Exchange Wang–Landau sampling method, we investigated and compared three different coarse-grained lattice protein models for the small, hydrophobic protein Crambin. We show that slight extensions of the HP lattice protein model, including the stiffness of bonds can lead to a significant decrease in ground-state degeneracies (up to 5 orders of magnitudes). Moreover, the ground-state structures begin to bear resemblance to native structures observed in real Crambin.

1. Introduction

Generic, coarse-grained protein models have served as great tools to improve our understanding of many important biological problems [1, 2]. Among these models, the hydrophobic-polar (HP) lattice protein model [3, 4] is arguably the simplest, yet it still remains computationally challenging. Finding the lowest energy structures of given sequences in the HP model has been proven to be an NP-complete problem [5], although different enumeration methods and stochastic algorithms have been developed for effectively approaching this problem. Examples include sequential importance sampling [6], fragment regrowth Monte Carlo [7], chain-growth methods [8, 9], e.g. the pruned-enriched Rosenbluth method (PERM) and its variants [10, 11, 12, 13, 14], multidomain sampler [15], genetic algorithms [16, 17], evolutionary Monte Carlo [18], ant colony models [19] and Wang–Landau sampling [20, 21]. With these latter methods, several problems of biological interest have been successfully investigated, such as the effects of sequence mutations [22, 23], protein folding in confined environments [24, 25] and protein surface adsorption [26, 27, 28, 29].

However, the simplicity of the HP model yields large ground-state degeneracies which stands in contrast to the generally unique native states of natural proteins. To overcome this issue, and also reflect more closely the range of hydrophobicities found among different amino acids a third "neutral" amino acid, noted by "0" has been indtroduced. The resultant H0P model generally shows less ground-state degeneracies compared to the original HP model [30]. In this paper we shall use Replica Exchange Wang-Landau (REWL) simulations to study the behavior of the protein Crambin mapped onto an H0P model. In addition, we take account of the natural rigidity of real proteins by introducing an energy term that reflects the cost of producing a bond angle between successive bonds. The purpose of this study is to investigate whether the introduction of these additional degrees of freedom may yield a significantly better representation of real proteins (here Crambin) while still remaining within the frame of minimalist models with all their advantages (e.g. rapid sampling of conformational space).

2. Models

2.1. H0P lattice protein model

Analogous to the HP model, the H0P model classifies amino acids into three types: hydrophobic (H), "neutral" (0) and polar (P), based on their hydrophobicity levels [31]. The primary structure of a protein is composed of a chain of connected (bonded) beads, each of which represents a single monomer, on a simple cubic lattice. Two non-bonded interactions are considered: $\epsilon_{\rm HH}$ and $\epsilon_{\rm H0}$, where $\epsilon_{\rm HH}$ denotes the interaction strength between non-bonded H-mers on nearest neighbor sites ($\epsilon_{\rm H0}$ has corresponding meaning), and $\epsilon_{\rm HH} > \epsilon_{\rm H0}$. The Hamiltonian is then given by

$$\mathcal{H} = -\epsilon_{\rm HH} n_{\rm HH} - \epsilon_{\rm H0} n_{\rm H0}, \tag{1}$$

where $n_{\rm HH}$ is the number of non-bonded HH contacts ($n_{\rm H0}$ has corresponding meaning). Other interactions involving "0" monomers could be easily added.

2.2. Semi-flexible H0P model

The idea of considering bond-stiffness energy has been studied previously, e.g., Thomas and Dill explored the relationship between helical propensities and conformations for globular proteins using an HP model with local helix interaction [32]; Bastolla and Grassberger [33] studied a lattice model of semi-flexible homopolymers with nearest neighbor attraction and energetic preference for straight joints between bonded monomers; Krawczyk et al. [34] have extensively investigated semi-flexible hydrogen-bonded and non-hydrogen bonded lattice polymers in both two and three dimensional lattices.

In this work, we include an energetic term ϵ_θ for each bent bond joint (i.e. angle) inside the protein structures resulting in the following Hamiltonian:

$$\mathcal{H} = -\epsilon_{\rm HH} n_{\rm HH} - \epsilon_{\rm H0} n_{\rm H0} - \epsilon_\theta n_\theta, \tag{2}$$

where n_θ represents the number of angles in a given protein structure. Figure 1 illustrates the various energy terms for a short semi-flexible H0P model.

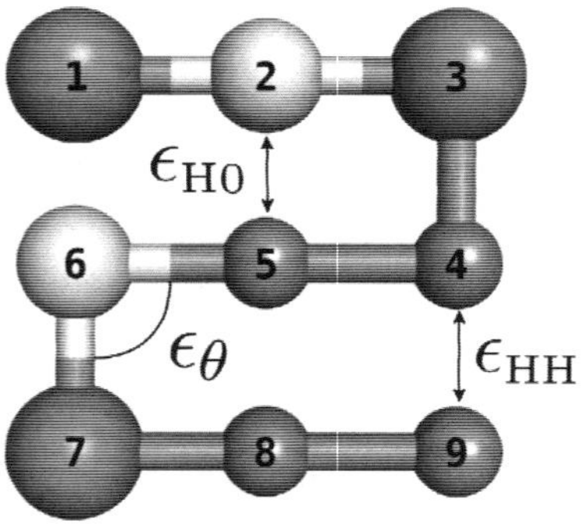

Figure 1. An example of a semi-flexible H0P model. Hydrophobic and "neutral" monomers are colored in gray and white, respectively, while polar monomers are colored in orange. The interaction between monomers 2 and 5 is $\epsilon_{\rm H0}$, and that between monomers 4 and 9 is $\epsilon_{\rm HH}$. The angle constituted by monomers 5, 6 and 7 contributes ϵ_θ energy. In this particular 2-dimensional structure, $n_{\rm HH} = 2$, $n_{\rm H0} = 1$ and $n_\theta = 4$.

3. Methods

3.1. Wang–Landau sampling

For systems with complex free energy landscapes, e.g., lattice proteins, traditional Monte Carlo methods such as Metropolis sampling are known to have trouble in sampling the low temperature region because of trapping in local minima. Wang–Landau (WL) sampling [35, 36] has the ultimate goal of estimating the density of states $g(E)$ iteratively by performing a random walk in a given energy range $[E_{\text{min}}, E_{\text{max}}]$ and effectively eliminates trapping. Together with suitable Monte Carlo trial moves, WL sampling has proven to be extremely powerful for studying this type of systems [20, 21]. During the simulation, each Monte Carlo trial move that changes the system from energy E_A to E_B will be accepted with probability

$$P(A \to B) = \min\left\{1, \frac{g'(E_A)}{g'(E_B)}\right\}, \tag{3}$$

where $g'(E)$ is an instantaneous estimator for the density of states. After each trial update $g'(E_i)$ is updated via $g'(E_i) \to f \times g'(E_i)$, where i refers to the state that results from the trial, f is a modification factor, and the histogram of visited energies is increased by one. The modification factor is reduced when the histogram is sufficiently "flat" and at the same time, all the entries in the histogram are reset to zero. The procedure is repeated until the modification factor is less than some predefined threshold value f_{final}. In this work, the initial modification factor is set to $f_{\text{init}} = e^1$ and decreased via $f \to \sqrt{f}$, and $f_{\text{final}} = 1 \times 10^{-8}$. Initially, all entries in $g'(E)$ are assigned to 1 and we apply the "80%" flatness criterion for the histogram, i.e., every entry in the histogram is no less than the 80% of the mean histogram height. As Monte Carlo trial moves, we have used pull and bond-rebridging moves which have been proven to be very efficient together with WL sampling [20, 21].

3.2. Replica-Exchange Wang–Landau sampling

Replica-Exchange Wang–Landau (REWL) [37, 38] sampling, a parallel version of Wang–Landau sampling, fully exploits and combines the power of WL sampling and replica-exchange Monte Carlo. It divides the energy range $[E_{\text{min}}, E_{\text{max}}]$ of the system into multiple, overlapping sub-ranges (windows). Configurations (replicas) in each of the resulting windows are simulated by independent processes (random walkers, running serial sampling, each on a different core), and then replica exchanges between overlapping windows are attempted at fixed time intervals during the simulation. As a result, each replica can travel through the entire energy space of interest. The probability of accepting a replica exchange move between random walker i and j from neighboring windows is given by

$$P_{\text{acc}} = \min\left\{1, \frac{g_i'(E_i)}{g_i'(E_j)}\frac{g_j'(E_j)}{g_j'(E_i)}\right\}, \tag{4}$$

where g_i and E_i denote the estimator for the density of states and the energy of current configuration of walker i, respectively. REWL exhibits very good strong and weak scaling properties for different systems including the Potts model, lattice protein adsorption, or the self-assembly process in amphiphilic solutions [37, 38] and is very effective in speeding up our simulations.

3.3. Ground state sampling

The problem of finding the ground-state structures of lattice protein models is NP-complete [5]. And due to the exponential growth of combinations, exact enumeration methods become unfeasible for protein sequences longer than 30 monomers. However, in previous work we proposed a ground state sampling method [23] that relies on Wang–Landau sampling to obtain reliable estimates of the ground-state degeneracies. The idea behind our method is to perform multicanonical sampling with the sampling weight as $1/\tilde{g}'(E)$, where $\tilde{g}'(E)$ is the final estimator obtained from the preceding WL or REWL runs. During the simulation, each encountered structure with lowest energy will be translated into a unique sequence of directions (SoD) containing five elements: Forward(F), Left(L), Right(R), Up(U) and Down(D), and will be stored in a tree data structure for efficient searching and comparison. When a ground-state structure is sampled, its SoD will first be compared with existing ones. Any newly discovered SoD will be added to the database. The whole process terminates when the ground-state degeneracy has "converged" (i.e., no further ground state is discovered for a period of time). Since a SoD uniquely represents a structure (excluding symmetry operations of the underlying simple cubic lattice), this algorithm provides an effective means of identifying and storing different ground-state structures and thus, estimating degeneracies.

4. Results and discussion

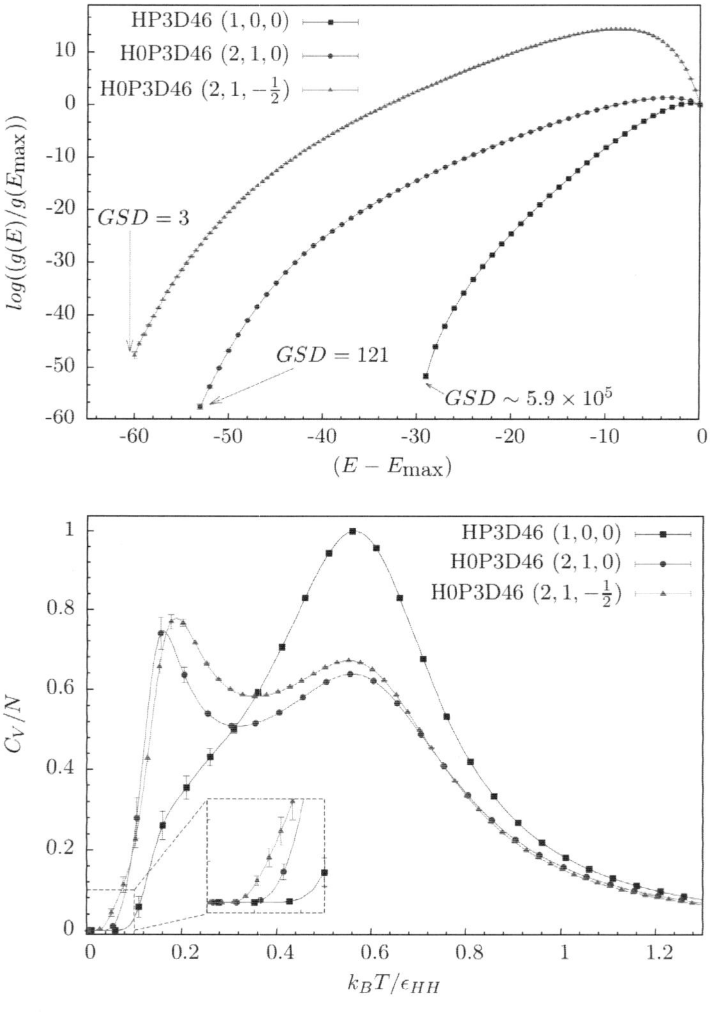

Figure 2. Observables from REWL simulations of HP3D46 and H0P3D46 lattice models for Crambin: (top) the densities of states with their ground-state degeneracies (GSD) and (bottom) specific heat (C_V/N) of HP3D46, H0P3D46 and semi-flexible H0P3D46. In both figures, error bars smaller than data points are not shown; tuples in the legend indicate the values of ($\epsilon_{\rm HH}$, $\epsilon_{\rm H0}$, ϵ_θ).

Crambin is a hydrophobic protein [39] with 46 residues. Its amino acid sequence has been converted to a HP sequence (denoted as HP3D46[1]) by Lattman et al. [40] and also to a H0P

[1] HP3D46: PPHHPPHHHPPPHPHHPHPPPPPHHHHPHPPHHHHPPHPHPPPHHP

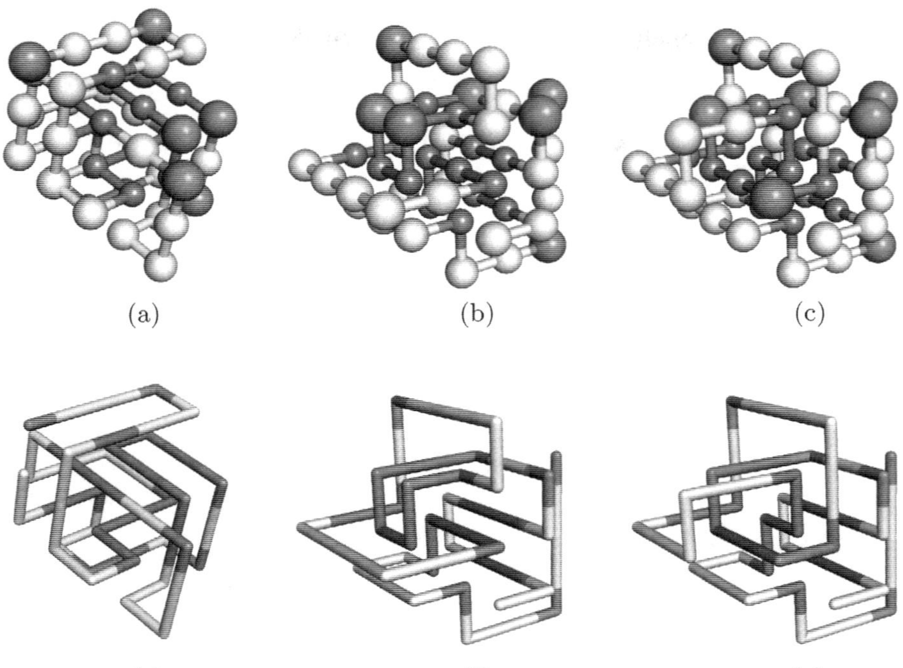

Figure 3. Top (a, b and c): Three ground-state structures of H0P3D46 with $\epsilon_{\rm HH} = 2$, $\epsilon_{\rm H0} = 1$ and $\epsilon_\theta = -0.5$. Hydrophobic and "neutral" monomers are colored in gray and white, respectively, while polar monomers are colored in orange. Bottom (e, f and g): Corresponding structures showing bonds only.

sequence (denoted as H0P3D46[2]) by Shi et al. [30]. It is interesting to note that the mapping onto an HP model yields a slightly polar protein whereas the H0P model mapping produces significantly more H-mers than P-mers. In this work, we studied both HP3D46 as well as H0P3D46 with and without the bond angle term. In this initial study we chose only one value for each kind of interaction (a more detailed study is in preparation). As shown in Figure 2, we present the densities of states as well as specific heats of HP3D46, H0P3D46 and semi-flexible H0P3D46. In Figure 2 (top), along with densities of states, we indicate the ground-state degeneracies (GSD) at the position of the lowest energy for each model. The GSD has decreased significantly from HP3D46 to H0P3D46 due to the addition of the "neutral" type of monomer and the (weaker) H-0 interaction. Equally significant is the decrease in ground state degeneracy for semi-flexible H0P3D46; only **three** nonequivalent ground-state structures, shown in Figure 3, were found. Because of the coarseness of the lattice model, the lattice protein structures could not perfectly resemble the protein structure in continuous space; however, one can observe from Figure 3 that each structure contains a "lattice helix" which agrees with the protein structure of Crambin determined by experiments [39]. Moreover, the specific heats (C_V/N) as shown in Figure 2 (b) indicate that, contrary to HP3D46, both H0P3D46 and semi-flexible H0P3D46 show pronounced two-step folding process signals as also observed in all-atom simulations [41]. The coil-globule or collapse transition occurs at the same temperature for all three lattice models, but the rearrangement at lower temperature (sometimes referred as the folding transition) clearly

[2] H0P3D46: 00HH00HH0P0PHPHHPH0000P0HH00000HHHH0000H00P00P

depends upon the details of the model.

Although the peak indicating the folding transition of semi-flexible H0P3D46 is at a slightly higher temperature than for the H0P model without the bond angle term, the specific heat rises more quickly at low temperature (as seen in the inset of Figure 2 (bottom)) due to the smaller energy gap between ground states and first excited states for semi-flexible H0P3D46 (as seen in Figure 2 (top)).

5. Conclusion

Mapping a real protein, Crambin, onto different lattice protein models, the HP model and the H0P model, and using the Replica-Exchange Wang–Landau sampling method, we were able to determine their thermodynamic properties as well as ground-state structures. We found that with the relatively minor generalization of the HP model , i.e. introducing the "neutral" monomer and bond stiffness (semi-flexible H0P model), we were able to reduce the ground-state degeneracies to extremely small values. Moreover, the ground-state structures obtained during the simulation contain features of a "lattice helix" structure which agrees with the experimental results.

Acknowledgments

This work was supported by NSF under Grant No. OCI-0904685; in part by resources and technical expertise from the Georgia Advanced Computing Resource Center, a partnership between the University of Georgias Office of the Vice President for Research and Office of the Vice President for Information Technology; part of the computing resources were provided by the Texas Advanced Computing Center under XSEDE Grant No. PHY130014.

References

[1] Dill K A 1999 *Protein Sci.* **8** 1166–80
[2] Kolinski A and Skolnick J 2004 *Polymer* **45** 511–524
[3] Dill K A 1985 *Biochemistry* **24** 1501–9
[4] Lau K F and Dill K A 1989 *Macromolecules* **22** 3986–3997
[5] Irbäck A and Troein C 2002 *J. Biol. Phys.* **28** 1–15
[6] Zhang J L and Liu J S 2002 *J. Chem. Phys.* **117** 3492
[7] Zhang J, Kou S C and Liu J S 2007 *J. Chem. Phys.* **126** 225101
[8] O'Toole E M and Panagiotopoulos A Z 1992 *J. Chem. Phys.* **97** 8644
[9] Beutler T C and Dill K A 1996 *Protein Sci.* **5** 2037–2043
[10] Grassberger P 1997 *Phys. Rev. E* **56** 3682–3693
[11] Frauenkron H, Bastolla U, Gerstner E, Grassberger P and Nadler W 1998 *Phys. Rev. Lett.* **80** 3149–3152
[12] Bachmann M and Janke W 2003 *Phys. Rev. Lett.* **91** 208105
[13] Hsu H P, Mehra V, Nadler W and Grassberger P 2003 *J. Chem. Phys.* **118** 444
[14] Prellberg T and Krawczyk J 2004 *Phys. Rev. Lett.* **92** 120602
[15] Tang W and Zhou Q 2012 *Phys. Rev. E* **86** 031909
[16] Unger R and Moult J 1993 *J. Mol. Biol.* **231** 75
[17] König R and Dandekar T 1999 *BioSystems* **50** 17–25
[18] Liang F and Wong W H 2001 *J. Chem. Phys.* **115** 3374
[19] Shmygelska A and Hoos H H 2005 *BMC Bioinf.* **6** 30
[20] Wüst T and Landau D P 2009 *Phys. Rev. Lett.* **102**(17) 178101
[21] Wüst T and Landau D P 2012 *J. Chem. Phys.* **137** 064903
[22] Holzgräfe C, Irbäck A and Troein C 2011 *J. Chem. Phys.* **135** 195101
[23] Shi G, Vogel T, Wüst T, Li Y W and Landau D P 2014 *Phys. Rev. E* **90** 033307
[24] Ping G, Yuan J M, Vallieres M, Dong H, Sun Z, Wei Y, Li F Y and Lin S H 2003 *J. Chem. Phys.* **118** 8042
[25] Pattanasiri B, Li Y W, Landau D P, Wüst T and Triampo W 2012 *J. Phys.: Conf. Ser.* **402** 012048
[26] Castells V, Yang S and Van Tassel P 2002 *Phys. Rev. E* **65** 031912
[27] Swetnam A and Allen M P 2012 *Phys. Rev. E* **85** 062901
[28] Radhakrishna M, Sharma S and Kumar S K 2012 *J. Chem. Phys.* **136** 114114

[29] Li Y W, Wüst T and Landau D P 2013 *Phys. Rev. E* **87** 012706
[30] Shi G, Wüst T, Li Y W and Landau D P 2015 *J. Phys.: Conf. Ser.* **640** 012017
[31] Kyte J and Doolittle R F 1982 *J. Mol. Biol.* **157** 105 – 132
[32] Thomas P D and Dill K A 1993 *Protein Science* **2** 2050–2065
[33] Bastolla U and Grassberger P 1997 *J. of Stat. Phys* **89** 1061
[34] Krawczyk J, Owczarek A and Prellberg T 2009 *Physica A* **388** 104–112
[35] Wang F and Landau D P 2001 *Phys. Rev. Lett.* **86** 2050–2053
[36] Wang F and Landau D P 2001 *Phys. Rev. E* **64** 1–16
[37] Vogel T, Li Y W, Wüst T and Landau D P 2013 *Phys. Rev. Lett.* **110** 210603
[38] Vogel T, Li Y W, Wüst T and Landau D P 2014 *Phys. Rev. E* **90** 023302
[39] Schmidt A, Teeter M, Weckert E and Lamzin V S 2011 *Acta Crystallographica Section F* **67** 424–428
[40] Lattman E E, Fiebig K M and Dill K A 1994 *Biochemistry* **33** 6158–6166
[41] Shimada J, Kussell E L and Shakhnovich E I 2001 *J. of Mol. Biol.* **308** 79 – 95

Homopolymer Adsorption on Hexagonal Surfaces: A Replica-Exchange Monte Carlo Study

B Liewehr[1] and M Bachmann[1,2,3]

[1]Soft Matter Systems Research Group, Center for Simulational Physics, The University of Georgia, Athens, Georgia, USA
[2]Instituto de Física, Universidade Federal de Mato Grosso, Cuiabá, Brazil
[3]Departamento de Física, Universidade Federal de Minas Gerais, Belo Horizonte, Brazil

E-mail: benjamin.liewehr@uga.edu, bachmann@smsyslab.org; http://www.smsyslab.org

Abstract. The adsorption behavior and thermodynamic properties of a coarse-grained flexible homopolymer, grafted on a hexagonal patterned surface, are investigated by means of parallel-tempering replica-exchange Monte Carlo simulations. In this study, the strength of the polymer-surface interaction, which is based on a standard Lennard-Jones potential, is changed systematically, mimicking different honeycomb-structured substrate materials. Utilizing specific order parameters, measured over a range of various temperatures and surface adsorption strengths, different structural phases are discriminated into classes of expanded, globular, droplet, and compact conformations. Finally, we construct the hyperphase diagram for a polymer with 55 monomers and discuss representative polymer structures.

1. Introduction

Recent research on two-dimensional nanomaterials revealed interesting surface properties, which are expected to lead to new applications in emerging technologies such as energy storage and novel electronic materials for integrated circuits. The most prominent example in this regard is graphene with outstanding mechanical, electrical, and thermal properties [1].

Since graphene as a semi-metal does not exhibit a band gap, which is required for many advanced applications like photovoltaics or field-effect transistors (FETs), structurally similar classes of materials have been investigated viz. transition metal dichalcogenides (TMX_2) and metal monochalcogenides (MX). Examples for these materials are MoS_2 or GaSe, exhibiting also hexagonal surface structures. These topological insulators and semiconductors allow for band gap engineering and thus show interesting optical properties such as photoluminescence. However, in contrast to pure graphene sheets, these materials are, in general, not chemically inert. Therefore, a protective coating is required to prevent reactions with the environment. Polymer coats are explored to serve this purpose [2, 3]. One key aspect of our study is the systematic investigation of polymer adsorption properties on generic hexagonal surfaces at different adsorption strengths, representing an entire class of miscellaneous substrates.

Graphene, as a potential substrate for drug delivery [1], may also be relevant for health care applications. Important in this context is the adhesion enhancement with polymers [4], as well as the compatibility with a biological environment [5, 6]. Therefore, it is necessary to understand the structural and thermodynamic behavior of such finite and comparatively small systems under thermal conditions. Since finite-size effects are essential for structure formation on mesoscopic scales, qualitative changes of macrostate ensembles of such systems are not phase transitions in the conventional sense, because the thermodynamic limit is out of reach and scaling properties are typically not of interest. Key towards understanding these small interacting systems is the systematic investigation of the cooperative polymer adsorption behavior, which is currently only accessible by means of computer simulations. The system-inherent complexity generally entails a rugged free-energy landscape in the space of relevant degrees of freedom, requiring advanced generalized-ensemble Monte Carlo sampling methods [7]. In order to unravel the general adsorption behavior of polymer chains on surfaces, we employ a generic hybrid coarse-grained polymer-substrate model, in which the monomers in the linear polymer chain, representing entire chemical groups, are attracted by the interaction sites of the patterned substrate. The polymer is grafted with one end to the surface [8–10].

Here, we present first insights into the adsorption behavior of a flexible polymer at hexagonal surfaces for different adsorption strengths and temperatures. Dominant structural conformations are identified in this parameter space and enable the construction of a structural hyperphase diagram.

2. Polymer-substrate model and evaluation of the surface potential

In this adsorption study, the interactions of a homopolymer with $N = 55$ monomers are modeled by a potential U_{NB} describing the van-der-Waals polarization effects between non-bonded monomers of the polymer, by a bond potential U_{B} between subsequent polymer beads, and by the interaction U_{S} of each bead with the vertices of a hexagonally patterned substrate.

The non-bonded interaction is given by a shifted and truncated 12-6 Lennard-Jones (LJ) potential,

$$U_{\mathrm{NB}}(r) = \begin{cases} U_{\mathrm{LJ}}(r) - U_{\mathrm{shift}}, & r < r_{\mathrm{c}}, \\ 0, & \text{otherwise}, \end{cases} \tag{1}$$

with

$$U_{\mathrm{LJ}}(r) = 4\epsilon\left[(\sigma/r)^{12} - (\sigma/r)^{6}\right], \tag{2}$$

where r is the monomer-monomer distance and σ parametrizes the van-der-Waals radius of a monomer. The potential shift $U_{\mathrm{shift}} = U_{\mathrm{LJ}}(r_{\mathrm{c}})$ is subtracted to avoid a discontinuity at the cutoff distance r_{c}. In our simulations, we fix the minimum location of the potential to $r_0 = 2^{1/6}\sigma \equiv 1$, the cutoff distance to $r_{\mathrm{c}} = 2.5\sigma$, and set the energy scale ϵ to unity.

The bonded potential between adjacent monomers is based on the finitely extensible nonlinear elastic (FENE) potential [11–13]. A Lennard-Jones potential is added to account for the asymmetry of the bond due to volume exclusion and polarization,

$$U_{\mathrm{B}}(r) = -\frac{1}{2}KR^2 \ln\left[1 - \left(\frac{r - r_0}{R}\right)^2\right] + \eta(U_{\mathrm{LJ}}(r) + 1) - 1 - U_{\mathrm{shift}}. \tag{3}$$

The FENE potential restricts the bond extension to the interval $r \in [r_0 - R, r_0 + R]$, with parameters fixed to the typical values $R \equiv 3/7$ and $k \equiv 90/4$. The parameter η is set to $\eta \equiv 0.1$. The constant shift leads to the same minimum value for U_{B} as for U_{NB}.[1] One terminal monomer of the homopolymer chain is grafted to a vertex of the hexagonal surface. While the positions of the remaining $(N - 1)$ monomers are sampled, locations of surface sites are kept fixed in the $z = 0$ plane, allowing for a three-fold rotational symmetry and two linearly independent translational symmetries. The interaction between a monomer and a lattice site at the distance r is given by

$$U_{\mathrm{S}}(r) = \epsilon_{\mathrm{S}}\, U_{\mathrm{NB}}(r), \tag{4}$$

[1] U_{B} can be shifted arbitrarily, because the number of bonds is kept constant. The shift leads only to an irrelevant overall energy constant.

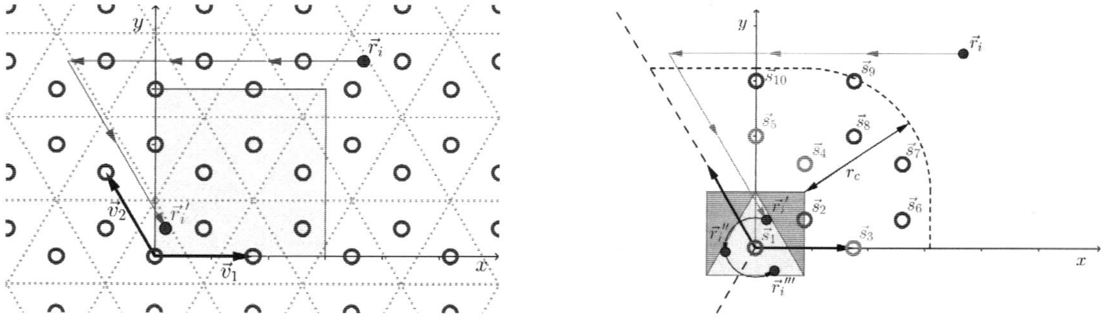

Figure 1. Hexagonal lattice structure of the substrate (circles). **Left**: Base vectors are denoted by $\vec{v}_1 = (\sqrt{3}, 0)^T$ and $\vec{v}_2 = (-\sqrt{3}/2, 3/2)^T$. The Wigner-Seitz cells are shown as dotted lines. If the monomer at position $\vec{r}_i$ is translated into the basis cell, its new location is $\vec{r}_i\,'$; rotation by $2\pi/3$ and by $4\pi/3$ yields $\vec{r}_i\,''$ and $\vec{r}_i\,'''$, respectively. **Right**: The gray rectangle marks the codomain of the lattice vector translation if applied to an arbitrary monomer position $\vec{r}_i$. The lattice sites located within the LJ-cutoff distance (dashed) are stored in sets of increasing distance to the codomain. The basis cell $\{\vec{s}_1, \vec{s}_2\}$ (gray) forms set 1, followed by set 2 $\{\vec{s}_3, \ldots, \vec{s}_5\}$ (orange), and set 3 $\{\vec{s}_6, \ldots, \vec{s}_{10}\}$ (brown).

where the parameter ϵ_S represents the surface adsorption strength. For simplicity, we choose the same cutoff radius r_c as for the non-bonded potential. In this study, the attraction strength of the surface was varied in the interval $\epsilon_\mathrm{S} \in [0.0, 4.5]$, effectively representing different materials or solvent properties. The total energy of a conformation $\mathbf{X} = (\vec{r}_1, \ldots, \vec{r}_N)$ with the monomer-monomer distance $r_{ij} = |\vec{r}_i - \vec{r}_j|$ and the distance $r_{ia} = |\vec{r}_i - \vec{s}_a|$ of a monomer i from a surface site a eventually reads,

$$E(\mathbf{X}) = \sum_{i=1}^{N-2} \sum_{j=i+2}^{N} U_\mathrm{NB}(r_{ij}) + \sum_{i=1}^{N-1} U_\mathrm{B}(r_{ii+1}) + \sum_{i=2}^{N} \sum_{a=1}^{\infty} U_\mathrm{S}(r_{ia}). \tag{5}$$

For the evaluation of the last term in Eq. (5), it is computationally favorable to store the ten vertex positions $\vec{s}_1, \ldots, \vec{s}_{10}$ of the basis cell, next-nearest cells, etc., in one $4\pi/3$-sector (see Fig. 1) by utilizing the following algorithm.

First, a monomer $\vec{r}_i$ is shifted by integer multiples of the lattice vectors towards the fundamental cell, where we assign to it the new position $\vec{r}_i\,'$. Second, two copies $\vec{r}_i\,''$ and $\vec{r}_i\,'''$ are created by rotation about the z-axis by $2\pi/3$ and $4\pi/3$, respectively. The calculation of the distances r_{ia} between monomers and surface sites is performed in sets of $\{\{\vec{s}_1, \vec{s}_2\}, \{\vec{s}_3, \ldots, \vec{s}_5\}, \{\vec{s}_6, \ldots, \vec{s}_{10}\}\}$ (see Fig. 1, right), where the positions $\vec{s}_1, \ldots, \vec{s}_{10}$ are sorted by increasing distance with respect to the origin of the coordinate system. The evaluation can be terminated as soon as all contributions from one set vanish. Finally, this procedure is applied to all monomers close to the surface, within LJ cutoff distance $z < r_\mathrm{c}$.

In Fig. 2(top left), the effective potential $U_\mathrm{S}(x, y, z)$ is plotted for $\epsilon_\mathrm{S} = 1$ as a function of the lateral coordinates x, y in the distance $z = 0.81$ above the surface, probing the monomer positions in a triangular overlayer. This overlayer is observed for adsorbed compact phases, where monomers in the surface plane are located inside the centers of hexagons, i.e., on the vertices of the Wigner-Seitz cells depicted in Fig. 1(left). A small lateral translation of a polymer bead from this favored position results in a steep increase in energy compared to a vertical displacement (Fig. 2, top right). Vertical cuts of the surface potential along $y = 1$ and $x = \sqrt{3}$ at $z = 0.81$ are shown in Fig. 2(bottom left and right, respectively). Note that

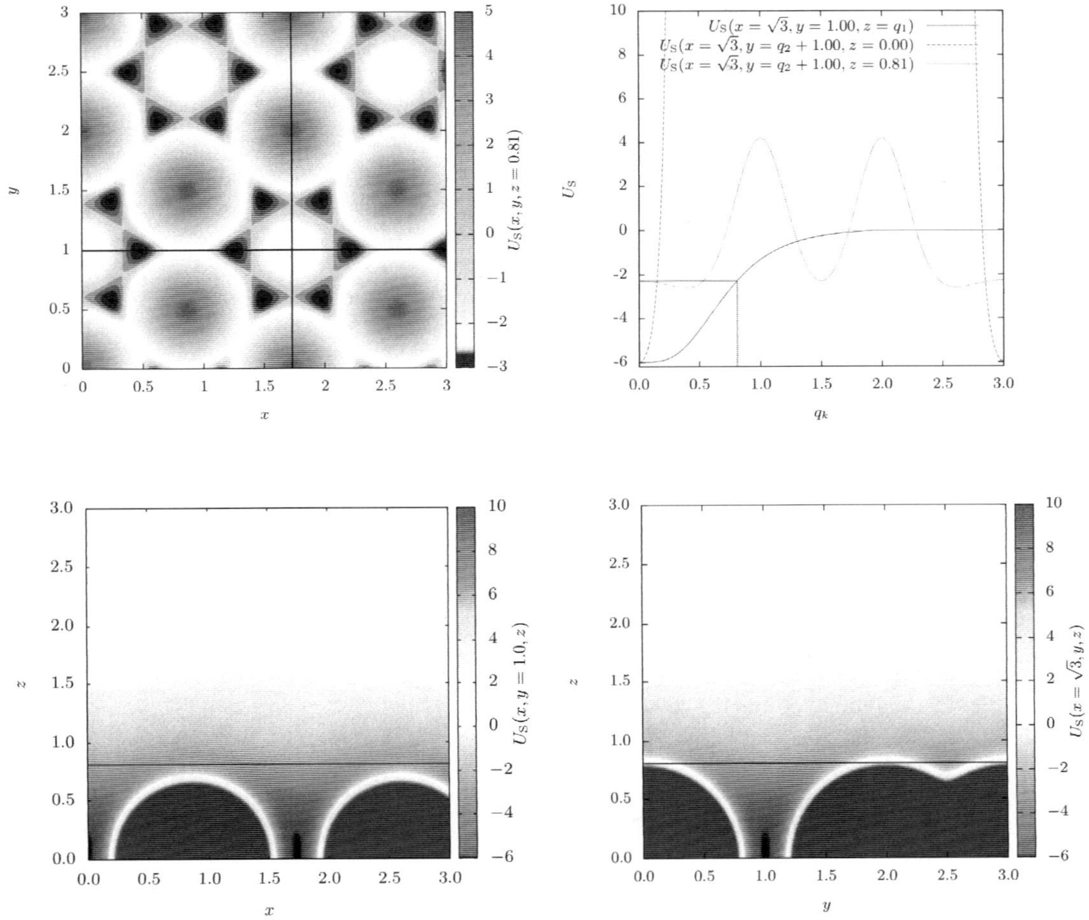

Figure 2. Representations of the surface potential $U_S(x, y, z)$ of the hexagonal substrate. **Top row**: (Left) Cut parallel to the surface at the height $z = 0.81$ of the triangular overlayer for the area marked in orange in the left-hand figure of Fig. 1. (Right) Potential as a function of favorable lattice site distance in perpendicular direction ($q_1 = z$, solid line), in a direction parallel to the surface ($q_2 = y - 1$), on the surface ($z = 0$, dashed), and above the surface ($z = 0.81$, dotted). **Bottom row**: Vertical cuts along $y = 1.0$ (left) and $x = \sqrt{3}$ (right). The height $z = 0.81$ is indicated by a horizontal line.

with increasing vertical distance the effective potential is smeared out. Thus, pattern effects are expected for polymer beads below $z = 1.5$ for this particular lattice type and parametrization of the potential.

3. Simulation method

The thermodynamic and structural properties of the 55-mer were investigated by means of data obtained in parallel tempering (replica exchange) Monte Carlo simulations at multiple temperatures, $T_1 < \ldots < T_m < \ldots < T_M$; $T_m \in [0.0, 5.0]$, uniformly distributed in inverse temperature space. Independent simulations were performed for a large number of surface attraction strengths $\epsilon_S \in [0.0, 4.5]$.

Whereas in each simulation the positions of surface sites were kept fixed, the conformational space of monomer positions was sampled at a given temperature T_m according to the Metropolis rule,

$$P_m(E_1 \to E_2) = \min\left(1, \exp\left[(E_1 - E_2)/k_B T_m\right]\right). \tag{6}$$

In every Monte Carlo sweep, a temperature-dependent displacement update was attempted for each non-grafted polymer bead. During the initial equilibration period, the maximum displacement was adjusted to allow for an acceptance rate of about 50%. Replica exchanges between temperature threads facilitated melting and freezing, i.e., decorrelation of polymer conformations. An exchange between neighboring temperature threads (T_m, T_k) was proposed after every 10^3 Monte Carlo sweeps and accepted with the exchange probability,

$$P_{mk}\left(E_m, T_m \leftrightarrow E_k, T_k\right) = \min\left(1, \exp\left[\left(E_m - E_k\right)\left(1/k_B T_m - 1/k_B T_k\right)\right]\right) \tag{7}$$

As the initial configuration in each thread we used the lowest-energy state found in a short test run. After thermodynamic equilibrium was reached, each simulation with fixed ϵ_S value comprised 10^5 replica exchange attempts per thread.

Thermodynamic properties were investigated by analyzing the canonical mean values such as energy $\langle E\rangle(T)$, components of the radius of gyration parallel and perpendicular to the surface, $\langle r_{\mathrm{gyr},\parallel}\rangle(T)$ and $\langle r_{\mathrm{gyr},\perp}\rangle(T)$, respectively, the end-to-end distance $\langle l_{\mathrm{ee}}\rangle(T)$, and number of contacts $\langle n_{\mathrm{s}}\rangle(T)$ at thread temperatures T_m.

Transition temperatures were identified by means of peaks and "shoulders" in fluctuation curves of a quantity O, given by

$$\frac{\mathrm{d}}{\mathrm{d}T}\langle O\rangle = \frac{\langle OE\rangle - \langle O\rangle\langle E\rangle}{k_B T^2}. \tag{8}$$

For example, the heat capacity $C_V = \mathrm{d}\langle E\rangle/\mathrm{d}T$ accounts for energetic fluctuations. Multiple-histogram reweighting [14] was employed to calculate the density of states $g(E)$ from individual estimates obtained in each temperature thread. The resolution of energy histograms was chosen as $\mathrm{d}E = 0.05$. With the density of states at hand, energetic quantities and fluctuations were calculated as continuous functions of temperature, which enabled us to locate transition signals between the simulation temperatures T_m more precisely.

Since the system is of finite size, transition signals identified in different response quantities for the same specific structural transition do not typically coincide. Therefore, this information was accumulated in T–ϵ_S parameter space in transition bands rather than transition lines. Phases separated by these bands were characterized by means of the polymer geometry, initially by analyzing similarity features of $10^2 - 10^3$ conformations within a single phase. This geometric characterization allowed for the introduction of phase-specific order parameters, which are described in the following section.

4. Adsorption on a hexagonal surface

4.1. Structural hyperphase diagram

A schematic overview of the adsorption behavior of the flexible homopolymer is presented in the T–ϵ_S hyperphase diagram shown in Fig. 3. As mentioned earlier, transition signals from different quantities are observed within a broad transition band, rather than at a unique transition temperature. Transition bands in Fig. 3 are indicated by gray lines, separating different structural phases which are annotated by a letter code, similar to the labels introduced in Ref. [9].

At high temperatures $T > 1.5$ and small adsorption strengths, the polymer resides primarily in the phase of desorbed expanded conformations (DE), which is well known from unconstrained

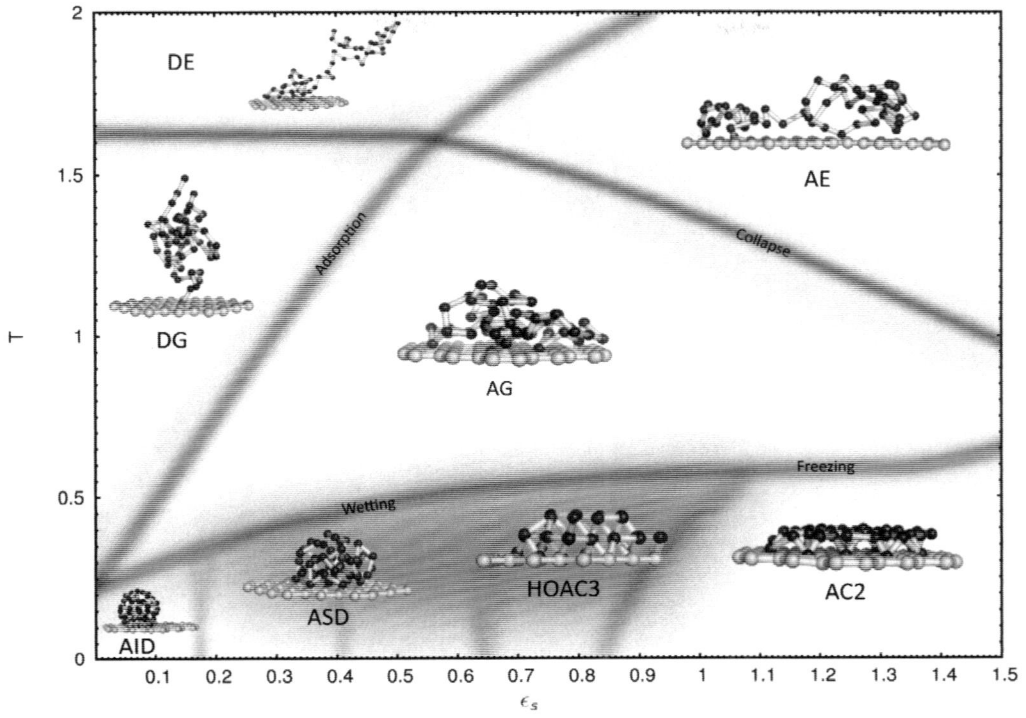

Figure 3. The structural T–ϵ_S hyperphase diagram is shown with typical conformations superimposed. At high temperatures and small adsorption strengths, the desorbed expanded (DE) conformations dominate. By crossing the adsorption transition line, the DE phase goes over into the adsorbed expanded (AE) phase for higher surface adsorption strength ϵ_S. Below the collapse transition band, adsorbed and desorbed globular phases are found. At low temperatures and depending on the surface attraction strength, adsorbed icosahedral droplets (AID) or semispherical droplets (ASD) form. For adhesion strengths $\epsilon_S > 0.8$, highly oriented adsorbed compact structures arranged in 3 layers (HOAC3) and adsorbed compact conformations (AC2) prevail.

homopolymers. For increased surface adsorption strength ϵ_S, these structures are found closer to the surface, within the Lennard-Jones cutoff $z < r_c$. Conformations in this regime are in energetic contact with the surface, but the pattern of the substrate is not yet recognized by the polymer. This polymer behavior defines the adsorbed expanded phase (AE). It is observed that the polymer beads in this phase are primarily located at rather large distances $1.5 < z < r_c$ where the surface potential is smeared out [see Fig. 2(top right)]. Therefore, the structural features in this phase compare well to the adsorption on continuous surfaces [15].

At low temperatures, generally compact polymer conformations govern the solid phases. Typical for a flexible homopolymer with 55 constituents is a crystalline structure with icosahedral shape. An icosahedron is characterized by 20 equally shaped tetrahedra. We find it attached to the surface in phase AID (adsorbed icosahedral droplet) if the surface absorption strength approaches zero. The shortest polymer chain forming such a structure is the 13-mer. Therefore, the 55-mer possesses an icosahedral core of 13 monomers while the remaining 42 monomers form the complete overlayer.

The more interesting and model parameter specific features are found for $\epsilon_S > 0$. Already for small surface interactions the symmetry of the icosahedral ground state is broken and adsorbed semi-spherical droplet-like (ASD) conformations dominate. At adsorption strengths $\epsilon_S > 0.8$ patterned structures are found. The number of surface contacts is a suitable order parameter that allows to distinguish the ordered phases, in which the surface pattern is recognized by the polymer, from disordered phases. It is defined as the number of monomers below a certain threshold distance $z_{\rm lim}$ to the surface and its fluctuation behavior is particularly useful for the identification of transition lines between these phases. One of the most interesting patterned subphases is HOAC3, which is dominated by highly oriented adsorbed compact 3-layer structures (see Fig. 3).

4.2. Thermodynamic and structural order parameters

One of the key challenges in this study is that surface-specific phases are found at very low temperatures, where parallel tempering generally suffers from large correlations that cause an effective slowing-down of the simulation. Moreover, structures in these phases depend sensitively on values of the surface adsorption strength ϵ_S. Consequently, we expect more subphases at low temperatures in the regime of droplet-like and patterned conformations [16]. Since signals in this regime are much weaker than for more prominent transitions, i.e., the freezing transition or coil-globule transitions, we suppose that subphases in the regime of droplet conformations are separated by steep (vertical) transition bands. In Fig. 3, these transitions are sketched in the interval $\epsilon_S \in [0.1, 1.1]$. With parallel tempering, they can only be identified if the ϵ_S-parameter space is scanned with high resolution and high accuracy of suitable order parameters. A useful parameter that describes the effective distance of the homopolymer from the surface is the z-component of the center of mass. In a similar way, fluctuations about the mean of the sum over all surface interactions, which is the last term in Eq. (5), signal these transitions as well.

More specific is the average number of surface contacts, i.e., the number of monomers located near the surface below $z_{\rm lim}$. In particular in the phase HOAC3 (cf. Fig. 3), the monomers that are arranged in three layers are well separated by the threshold values $z_{\rm lim} \in \{0.3, 1.0, 1.5\}$. The latter value, for instance, is used to signal the freezing transition from the adsorbed globule (AG) into the highly oriented adsorbed compact (HOAC3) phase, since the number of surface contacts $n_{{\rm s},1.5}$ discriminates monomers that reside above the ordered layers which do not belong to the crystalline conformation.

In addition, we consider the threshold for an energetic contact with the surface given by the LJ cutoff $r_c \approx 2.23$. For small surface adsorption strengths, the fluctuation signals of $n_{{\rm s},0.3}$ compare well with the peaks in the heat capacity C_V. Furthermore, fluctuations of $n_{{\rm s},1.0}$ suggest vertical separations of droplet and globular subphases. For increasing surface adsorption strengths, polymer conformations become flatter (wetting). The divergence of the FENE potential at maximum bond length prevents the occupation of favorable lattice sites by two adjacent monomers of the homopolymer. Thus, at least one additional monomer is required to link monomers residing in the centers of neighboring hexagons. In the simplest case, three monomers form a *zig-zag* bridge, with the central monomer located above the surface. However, the repulsive character of the surface potential near the interaction sites of the substrate, as it is felt by the central monomer (Fig. 2, bottom row), requires that either monomers in the surface layer slightly lift off from the hexagonal plane or that bonds are stretched. Therefore, the formation of *zig-zag* bridges is strongly dependent on the surface adsorption strength. In the AC2 phase, *zig-zag* bridges can occur at the edge of the film-like structure.

Towards the center of typical AC2 conformations ("bulk"), beads in the second layer ($z \approx 0.81$) form a regular triangular lattice. The bond length of bonded monomers located in two different layers typically corresponds to the equilibrium distance r_0. Obviously, the formation of the triangular overlayer requires more beads than are necessary to fill favorable lattice sites in the underlying surface layer. In this phase, we observe a ratio $(\langle n_{{\rm s},1.0}\rangle - \langle n_{{\rm s},0.3}\rangle)/\langle n_{{\rm s},0.3}\rangle \approx 3$. For higher surface adsorption strengths, $\epsilon_S > 2$, the formation of a triangular overlayer is reduced in favor of *zig-zag* bridges, leading to a larger number of surface contacts $n_{s,0.3}$.

5. Conclusion

In this work, we have studied the cooperative adsorption behavior of a generic polymer model interacting with a hexagonally patterned substrate by means of parallel-tempering replica-exchange Monte Carlo simulations. The interaction strength between the homopolymer and the surface was varied systematically.

To locate structural phase transition bands, we extracted canonical and structural quantities from sampled configurations and applied canonical statistical analysis. Different structural phases were characterized, allowing for the identification of pattern recognition effects by the homopolymer. We summarized these results by constructing a complete hyperphase diagram, parametrized by the temperature T and the surface adsorption strength ϵ_S. Depending on the adsorption strength, we found for low temperatures $T < 0.5$ specific wetting and freezing transitions into phases dominated by adsorbed droplet-like and crystal-like conformations. For high temperatures, $T > 1$, the entropic enhancement prevents pattern recognition by the polymer. In the phases of globular and extended structures, most monomers reside in a distance from the surface where patterns in the potential are averaged out.

The hyperphase diagram of adsorption for a flexible polymer near a hexagonal substrate exhibits a large number of characteristic and unique features. Depending on the surface adsorption strength and temperature, we identified different adsorption scenarios for surface wetting/dewetting and pattern recognition. Prospectively, this allows for both, the classification of surfaces with respect to polymer adsorption and it helps understand how different polymers adsorb on a certain substrate of interest. These questions are apparently of interest for the design of hybrid materials composed of soft and solid matter.

Acknowledgments

B. Liewehr acknowledges support by the *Studienstiftung des deutschen Volkes* and through the exchange program between the University of Rostock and the University of Georgia. This work has been supported partially by the NSF under Grant No. DMR-1207437 and by CNPq (National Council for Scientific and Technological Development, Brazil) under Grant No. 402091/2012-4.

References

[1] Novoselov K S, Fal'ko V I, Colombo L, Gellert P R, Schwab M G and Kim K 2012 *Nature* **490** 192–200
[2] Potts J R, Dreyer D R, Bielawski C W and Ruoff R S 2011 *Polymer* **52** 5–25
[3] Alexandre M and Dubois P 2000 *Materials Science and Engineering: R: Reports* **28** 1–63
[4] Sides S W, Grest G S and Stevens M J 2001 *Physical Review E* **64** 050802
[5] Wei Q, Becherer T, Angioletti-Uberti S, Dzubiella J, Wischke C, Neffe A T, Lendlein A, Ballauff M and Haag R 2014 *Angewandte Chemie International Edition* **53** 8004–8031
[6] Carignano M and Szleifer I 2000 *Colloids and Surfaces B: Biointerfaces* **18** 169–182
[7] Bachmann M 2014 *Thermodynamics and Statistical Mechanics of Macromolecular Systems* (Cambridge University Press, Cambridge)
[8] Li T and Park K 2001 *Computational and Theoretical Polymer Science* **11** 133–142
[9] Möddel M, Janke W and Bachmann M 2011 *Macromolecules* **44** 9013–9019
[10] Möddel M, Janke W and Bachmann M 2014 *Physical Review Letters* **112** 148303
[11] Bird R B, Curtiss C F, Armstrong R C and Hassager O 1987 *Dynamics of Polymeric Liquids* 2nd ed (Wiley, New York)
[12] Kremer K and Grest G S 1990 *Journal of Chemical Physics* **92** 5057–5086
[13] Milchev A, Bhattacharaya A and Binder K 2001 *Macromolecules* **34** 1881–1893
[14] Ferrenberg A M and Swendsen R H 1989 *Physical Review Letters* **63** 1195–1198
[15] Pattanasiri B, Liewehr B and Bachmann M 2015 *Physics Procedia* **68** 90–94
[16] Liewehr B and Bachmann M 2015 *to be published*

Journal of Physics: Conference Series **686** (2016) 012003 doi:10.1088/1742-6596/686/1/012003

Chain-growth simulations of the HP model for proteins

Fabricio S H Santos[1], **Paulo H L Martins**[1], **and Michael Bachmann**[1,2,3]

[1] Instituto de Física, Universidade Federal de Mato Grosso, Av. Fernando Corrêa da Costa, 2367, 78060-900, Cuiabá, MT, Brazil

[2] Soft Matter Systems Research Group, Center for Simulational Physics, Department of Physics and Astronomy, The University of Georgia, Athens, Georgia 30602, USA

[3] Departamento de Física, Instituto de Ciências Exatas, Universidade Federal de Minas Gerais, C.P. 702, 30123-970, Belo Horizonte, MG, Brazil

E-mail: fabricio@fisica.ufmt.br, pmartins@fisica.ufmt.br, bachmann@smsyslab.org

Abstract. We consider a generalized hydrophobic-polar model for proteins on square lattices. Besides the attraction between nonbonded hydrophobic monomers, the present model also takes into account an interaction between hydrophobic and polar units. By using the pruned-enriched Rosenbluth method (PERM), we investigate a specific polymer sequence composed of 42 monomers that has been proposed to simulate the physical properties of the parallel β-helix of pectate lyase C. For each temperature, the total number of generated chains varies from 10^6 to 10^7. Physical observables such as specific heat, total energy, end-to-end distance, radius of gyration, and the average number of hydrophobic-hydrophobic and hydrophobic-polar contacts are evaluated for different values of the ratio between the hydrophobic-hydrophobic and the hydrophobic-polar contact energies. Eventually, a pseudo-phase diagram in the space of temperature and the ratio of contact energy scales is constructed.

1. Introduction

Proteins are responsible for a wide range of biological functions, including transport and storage of molecules, coordinated movements, enzymatic catalysis, mechanical support, and others [1]. A protein is a linear molecule composed of a particular sequence of monomer units (amino acids) [2]. Since there exist 20 different amino acids in nature, a certain protein with N monomers is only one among 20^N possible polypeptides with this chain length [3]. The complexity of the system arises from this huge number of combinations. Due to the enormous number of tasks to be necessarily fulfilled to ensure the stability of a biological system, a large number of proteins exists [4], although it is vanishingly small compared to the number of possible sequences.

The biological function of a protein chain is closely related to its sequence and, hence, to its native conformation [5]. In this sense, it is very useful to identify the structure which a specific sequence will assume according to its environmental conditions. It is well known that, as the temperature increases, a heteropolymer chain undergoes changes in its shape, from the native fold to globules, and random coils [6].

In this work, we study a simple two-dimensional lattice model for proteins and systematically investigate the dependence of the structural behavior on the environmental temperature and the ratio of energy scales associated with contacts between nonbonded hydrophobic monomers and mixed hydrophobic-polar contacts. For this purpose, we employ a chain-growth algorithm for the simulation of this system. We calculate quantities like the specific heat, mean energy, end-to-end distance, and radius of gyration. Different temperatures and solvent conditions are considered to determine the phase diagram of a specific sequence composed by 42 monomers [7, 8, 9].

2. Model and Method

Despite the variety of monomer units, simplified models have been proposed in order to better understand the conformational transitions of a polymer chain. A linear chain of identical monomers is called a homopolymer, whereas heteropolymers are composed of different units [7, 10].

The hydrophobic-polar (HP) model is one of the simplest lattice models used to describe, very qualitatively, the folding behavior of proteins [7]. It considers two types of monomers, hydrophobic (H) and polar (P), and the conformations are modeled as interacting self-avoiding walks on the lattice [11]. It also assumes that the hydrophobic interaction is the essential driving force towards the native fold [12]. The chain studied here is a 42mer with the sequence: $PH_2PHPH_2PHPHP_2H_3PHPH_2PHPH_3P_2HPHPH_2PHPH_2P$, which was designed to simulate some properties of the parallel β-helix of *pectate lyase C* [13]. The monomers can occupy sites on a two-dimensional square lattice [14]. Adjacent monomers in the polymer sequence have a fixed unity bond length (one lattice unit). Detailed studies of this sequence on three-dimensional lattices, including only hydrophobic-hydrophobic (HH) interactions, have been performed in the past (see, e.g., Ref. [4] and references therein). Here, besides the HH attractive interaction, we also consider a hydrophobic-polar (HP) attraction. Therefore, the total energy of the system reads:

$$E = -\epsilon_{\rm HH} n_{\rm HH} - \epsilon_{\rm HP} n_{\rm HP} \ , \tag{1}$$

where $n_{\rm HH}$ and $n_{\rm HP}$ are the total number of non-bonded HH and HP pairs of neighboring monomers, respectively. Defining $s \equiv \epsilon_{\rm HP}/\epsilon_{\rm HH}$ and setting $\epsilon_{\rm HH} = 1$, one has

$$E = -(n_{\rm HH} + s\, n_{\rm HP}) \ . \tag{2}$$

Note that the limit $s = 0$ corresponds to the original HP model, with HH contacts only. The case $s = 1$ means that HH and HP contacts have equal strength.

We have performed chain-growth simulations based on the *pruned-enriched Rosenbluth method* (PERM) [6]. This algorithm has proven to be particularly useful for studying the θ collapse transition between random-coil and globular structures of homopolymers, but it has also been successfully applied for lattice heteropolymers. It uses the *go with the winners* strategy in the sense that it favors chains with small entropy as reflected by large Rosenbluth weights, which in PERM are introduced as [6]:

$$W_n^{PERM} = \prod_{l=2}^{n} m_l \exp^{-(E_l - E_{l-1})/k_B T} \ . \tag{3}$$

Here, n is the total length of the chain, E_l is the energy of the partial chain with size l, k_B is the Boltzmann constant, and T is the temperature. The average value of a physical observable Q can be calculated through

$$\langle Q \rangle = \lim_{M \to \infty} \frac{\sum_{\alpha=1}^{M} Q_\alpha W_{n,\alpha}^{PERM}}{\sum_{\alpha=1}^{M} W_{n,\alpha}^{PERM}} \ , \tag{4}$$

where M is the number of generated chains, Q_α is the value of quantity Q for the chain α, and $W_{n,\alpha}^{PERM}$ is the weight associated with that chain. For the subsequent discussion of structural phases in our model, the most interesting quantities are the specific heat and the average number of hydrophobic-hydrophobic and hydrophobic-polar contacts.

3. Results

Considering the HP chain embedded in an implicit solvent, we have simulated and analyzed the structural behavior of the heteropolymer for several values of HP/HH energy scale ratio s in the range $0 \leq s \leq 1$. Temperatures ranged from 0.1 to 2.0. For given (s, T) parameters, the total number of generated chains varied from 10^6 to 10^7. Figure 1 shows, on the left side, the specific heat C_V as a function of temperature T for selected values of s. One can note that, for $s = 0.0$, a single peak in the low-temperature region is observed, while for $s \geq 0.4$, two distinct peaks appear, corresponding to two conformational transitions. The mean energy $\langle E \rangle$ is plotted for the same s values in Fig. 1 (right). As expected, as the temperature increases, the energy approaches zero, indicating that the system has reached the random coil phase.

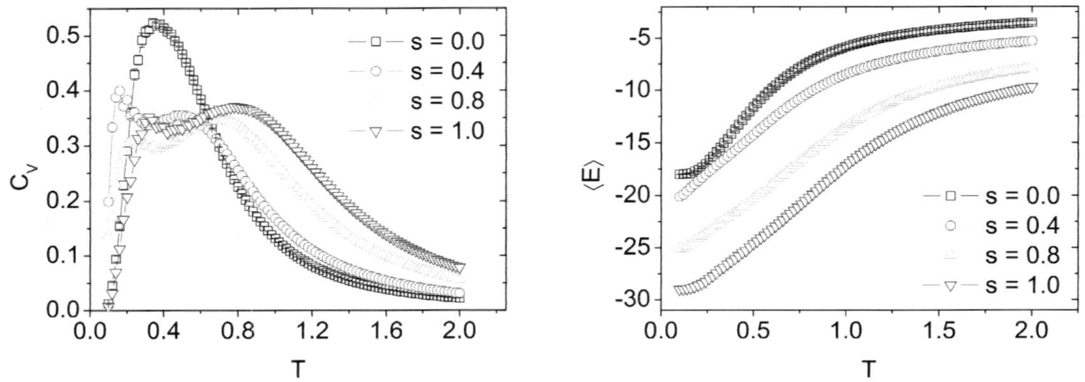

Figure 1. Specific heat (*left*) and mean energy (*right*) as functions of temperature for selected values of the parameter s. Error bars, if not visible, are smaller than the symbols.

In Fig. 2, we show the average number of HH and HP contacts (n_{HH} and n_{HP}, respectively). For small values of s (0.0 and 0.4, for instance), one can observe that the low-temperature conformations possess a significant number of HH contacts, while the number of HP contacts is small. As the parameter s approaches 1.0 (this limit means that HH and HP interactions are equal), n_{HP} gets larger while n_{HH} decreases.

To have an idea of how conformations look like under different thermal conditions, it is helpful to analyze structural quantities such as end-to-end distance, R_{ee}, and radius of gyration, R_g. Mean values of both quantities are shown in Fig. 3. An interesting behavior can be noticed for $s = 0.0$. In this case, the lowest energy we have found is -19. In contrast with the three-dimensional case [15], the 2D ground state seems to be highly degenerated with rather large variations of compactness. To maximize n_{HH} (and consequently minimize the total energy), the polymer forms an H-core that also includes some P monomers in a compact shape. Other ground-state conformations possess a hydrophobic backbone surrounded by polar monomers, but these structures have a much larger radius of gyration. As the temperature increases, $\langle R_{ee} \rangle$ and $\langle R_g \rangle$ decrease and reach a minimum value at $T \approx 0.25$. For higher values of s (especially for $s \geq 0.3$), as the HP contacts become more relevant, this behavior is not observed.

Figure 4 summarizes our results. It depicts the phase diagram in the $s - T$ space and includes some typical conformations in each region of the phase diagram. The transition points were obtained from the peak positions of the specific heat. In the low-temperature region, we find compact conformations dominating for large values of s, whereas conformations are less

Journal of Physics: Conference Series **686** (2016) 012003 doi:10.1088/1742-6596/686/1/012003

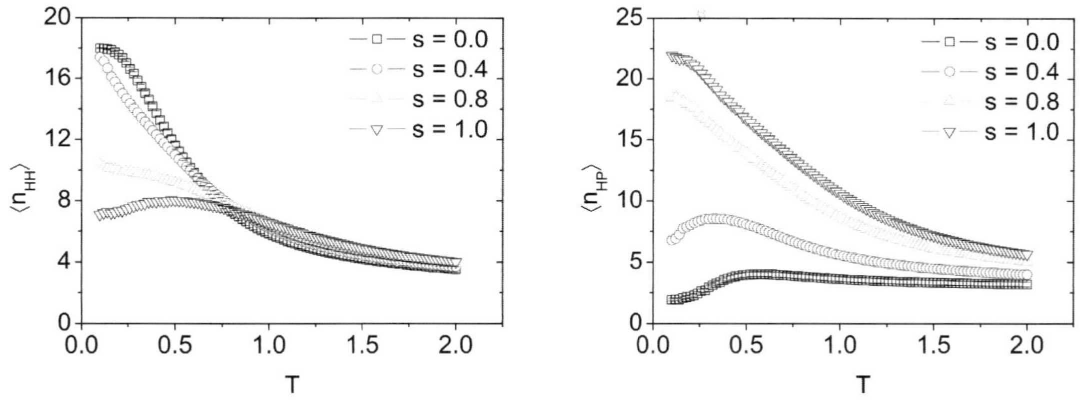

Figure 2. *Left*: average number of hydrophobic-hydrophobic contacts $\langle n_{HH} \rangle$. *Right*: average number of hydrophobic-polar contacts $\langle n_{HP} \rangle$. Error bars are of the order of the symbols size.

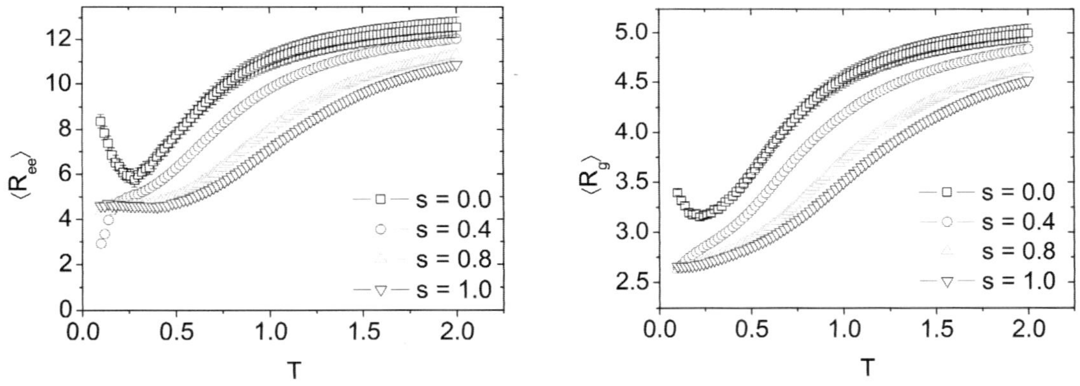

Figure 3. Average values of the end-to-end distance (*left*) and the radius of gyration (*right*). For better visualization, error bars are shown for $s = 0.0$ only. For the other values of s, error bars have similar sizes.

compact for small values of s, in which case HH contacts are favored. For a fixed value of s ($s \lesssim 0.20$) and increasing temperature, random coil structures form at $T \approx 0.38$, characterized by large values of the end-to-end distance and the radius of gyration. For $s \gtrsim 0.25$, there is an intermediate globular phase, in which both HH and HP contacts are relevant.

4. Conclusion

By using the pruned-enriched Rosenbluth method (PERM), we have analyzed a generalized version of the HP model for proteins on square lattices, including both hydrophobic-hydrophobic and hydrophobic-polar interactions. Several values of s (the strength of the HP attraction, relative to the HH interaction) were considered and results compared. The maximum values of the specific heat of a particular 42mer were used to determine the pseudophase diagram in the

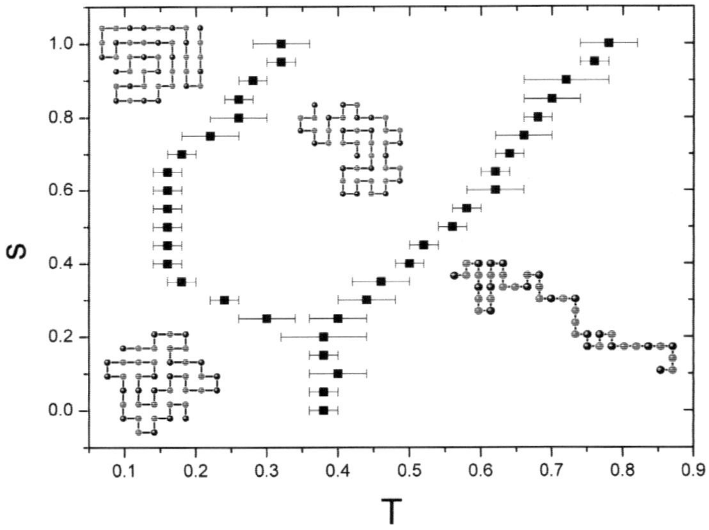

Figure 4. Pseudo-phase diagram in $s - T$ space. In the conformations shown, an H monomer is represented by a red bead while P monomers are black.

$s - T$ plane. The most interesting result of our study is that for HP/HH energy scale ratios $s \geq 0.2$, an intermediate phase of compact, but less ordered conformations forms ("liquid"). Studies of the present model in three dimensions are in progress.

Acknowledgments

Financial support was provided by FAPEMAT, CAPES, and CNPq (Grant No. 402091/2012-4). MB also acknowledges support by the NSF under Grant No. DMR-1207437.

References

[1] Stryer L 1995 *Biochemistry* (New York: W. H. Freeman & Company)
[2] Lau K F and Dill K A 1989 *Macromolecules* **22** 3986
[3] Dill K A, Bromberg S, Yue K, Fiebig K M, Yee D P, Thomas P D, and Chan H S 1989 *Protein Sci.* **4** 561
[4] Bachmann M 2014 *Thermodynamics and Statistical Mechanics of Macromolecular Systems* (Cambridge: Cambridge University Press)
[5] Branden C and Tooze J 1991 *Introduction to Protein Structure* (New York: Garland)
[6] Grassberger P 1997 *Phys. Rev. E* **56** 3682
[7] Dill K A 1985 *Biochemistry* **24** 1501
[8] Bachmann M and Janke W 2003 *Phys. Rev. Lett.* **91** 208105
[9] Bachmann M and Janke W 2004 *J. Chem. Phys.* **120** 6779
[10] Flory P J 1971 *Principles of Polymer Chemistry* (Ithaca: Cornell University Press)
[11] Chan H S and Dill K A 1993 *Phys. Today* **46** 24
[12] Dill K A 1990 *Biochemistry* **29** 31
[13] Yoder M D, Keen N T, and Jurnak F 1993 *Science* **260** 1503
[14] Orr W J C 1947 *Trans. Faraday Soc.* **43** 12
[15] Bachmann M and Janke W 2003 *Acta Phys. Pol. B* **34** 4689

Journal of Physics: Conference Series **686** (2016) 012004 doi:10.1088/1742-6596/686/1/012004

Structural behavior of an anomalous fluid under hydrophobic, hydrophilic and heterogeneous confinement

Rafael de C. Barbosa[1], **Leandro B. Krott**[2] **and Marcia C. Barbosa**[1]

[1] Instituto de Física, Universidade Federal do Rio Grande do Sul, Caixa Postal 15051, 91501-970, Porto Alegre, RS, Brazil

[2] Centro Araranguá, Universidade Federal de Santa Catarina, Rua Pedro João Pereira, 150, 88900-000, Araranguá, SC, Brazil

Abstract.
We perform molecular dynamics simulations of an anomalous fluid confined between parallel plates. The effects of the confinement on the structural behavior of the fluid when the plates are hydrophobic, hydrophilic and amphiphilic were investigate. The anomalous fluid was modeled using a two length scales potential, that exhibits density, diffusion and structural anomalies like observed in liquid water. We show that this fluid at the contact layer exhibits the formation of a solid/amorphous phase for temperatures at which the bulk system is liquid.

1. Introduction

Anomalous fluids, like water [1] and silica [2], exhibit a region in the pressure-temperature phase diagram at which thermodynamic, dynamic and structural properties behave differently when compared with other fluids. One example is the density of water that has a maximum value at $T = 4^oC$ at room pressure. Another example is its diffusion increases as the density decreases in a certain range of temperatures. But not only bulk water shows anomalies. The behavior of confined water is also very intriguing. In principle, under confinement, the disordered imposed by the confined walls should simply shift to lower temperatures all the bulk properties. Such a simple picture is not valid for water. At the nano scale for instance the fluid-wall hydrophobic/hydrophilic interactions that in the bulk are irrelevant compete with the fluid-fluid interactions [3, 4, 5] and lead to layering [6], large increase in the diffusion coefficient [6] and superflow [7, 8]. At the confining surface the water forms new ordered phases not observed in the bulk [9, 10, 11, 12].

Experiments of confined fluids show additional properties not observed in bulk systems. NMR and X-ray diffraction studies show that the pore size affects the freezing and the melting temperatures [13, 14]. But these results depend on the type of interaction between fluid and confining media. For the hydrophobic confinement the fluid remains at the liquid state for temperatures lower than in the bulk or for the hydrophilic confinement [3, 4, 5]. The freezing in both cases is not uniform. The contact layer and the middle layer do not melt at the same temperatures [15]. Therefore, a simple picture of the mechanism for the behavior of anomalous liquids under confinement is still lacking.

In addition to the atomistic models, core-softened potentials have been employed to describe the anomalous behavior of water-like systems [16, 17, 18, 19] in the bulk and in planar surfaces [20, 21, 22, 23]. These models represent the van der Waals and the hydrogen bonds interactions of water by means of a two length scales potential. The competition between the two scales generates in the bulk the density, diffusion and structural anomalies present in water. Even thought the effective potentials can not produce quantitative agreement with the experiments they provide a picture of the mechanism behind the anomalies.

In this work we study the behavior of an anomalous liquid under confinement in the framework of effective interaction potentials. The fluid is modeled by a two length scales potential [17, 24, 23, 25]. The plates are immersed in the fluid. Three systems are analyzed: hydrophobic, hydrophilic and heterogeneous plate-fluid interactions. These three cases are analyzed for the presence of freezing and melting.

The paper is organized as follow: in the section 2 the model is introduced, in the section 3 the methods and simulations details are described, in the section 4 the results are shown and in the section 5 conclusions are presented.

2. The Model

The system is composed of two parallel plates immersed in a fluid of N spherical particles of diameter σ. The plates are made of particles of diameter σ organized in a square lattice of area L^2. The distance between the parallel plates is given by d. A schematic representation of the system is depicted in the figure 1.

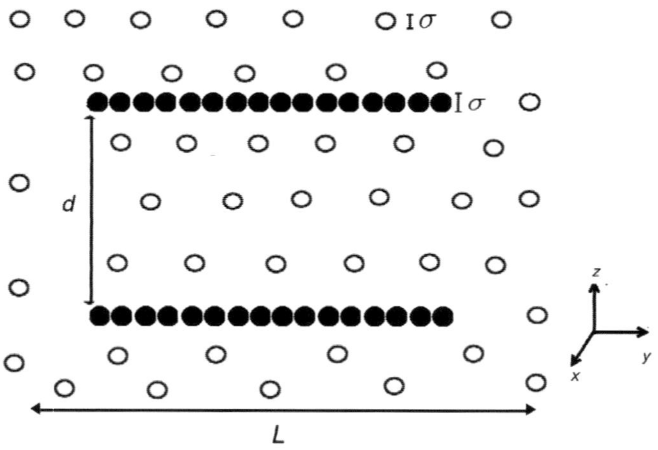

Figure 1. Schematic representation of the system formed by plates immersed in the fluid of spherical particles.

The particles of the fluid interact through a two length scales potential given by

$$\frac{U(r_{ij})}{\varepsilon} = 4\left[\left(\frac{\sigma}{r_{ij}}\right)^{12} - \left(\frac{\sigma}{r_{ij}}\right)^{6}\right] + a\exp\left[-\frac{1}{c^2}\left(\frac{r_{ij}-r_0}{\sigma}\right)^2\right] \tag{1}$$

where $r_{ij} = |\vec{r}_i - \vec{r}_j|$ is the distance between two particles, i and j. The first term is a standard Lennard-Jones (LJ) 12-6 potential where ε is the depth, plus a gaussian centered on radius r_0 and width c. The chosen parameters are $a = 5$, $r_0/\sigma = 0.7$ and $c = 1$. Oliveira et. al [17, 24] studied the pressure versus temperature phase diagram of the bulk system interacting through this potential and they found that the system exhibits a region in the pressure-temperature phase diagram where the density and the diffusion anomalous behavior are present. The presence of

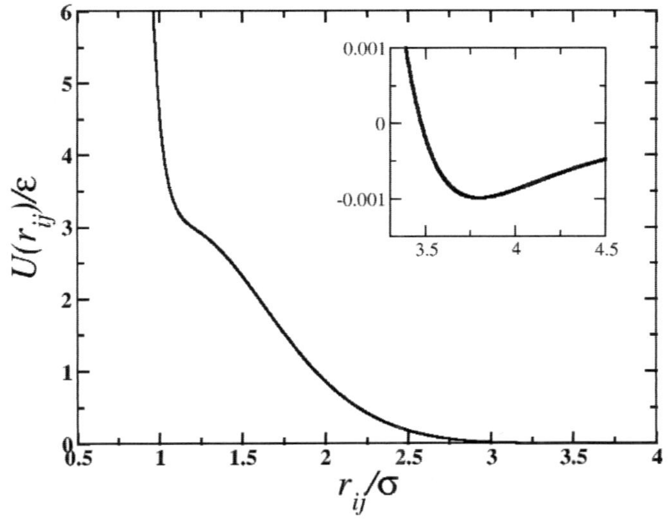

Figure 2. Effective potential of interaction between the water-like particles (Eq. 1). The potential parameters are $a = 5$, $r_0/\sigma = 0.7$ and $c = 1$. The inset shows a zoom of the small attractive part of the potential.

the anomalies is related with the two length scales: a repulsive shoulder at $r_{ij}/\sigma \approx 1$ and a very small attractive well at $r_{ij}/\sigma \approx 3.8$, as it can be seen in the figure 1. The same set of parameters is employed in this paper.

In order to check the effect of the particle-plates interaction potential in the behavior of the confined fluid three different types of walls are studied: hydrophobic, hydrophilic and heterogeneous fluid-wall interactions.

The hydrophobic interaction was simulated by the Weeks-Chandler-Andersen Lennard-Jones potential [26], given by

$$\frac{U_{WCA}}{\varepsilon} = \begin{cases} U_{\rm LJ}(r_{ij}) - U_{\rm LJ}(r_{cw}), & r_{ij} \leq r_{cw} \\ 0, & r_{ij} > r_{cw} \,, \end{cases} \tag{2}$$

where $U_{LJ}(r_{ij})$ is a standard 12-6 Lennard-Jones potential and $r_{cw} = 2^{1/6}\sigma$ is the cutoff distance.

The hydrophilic interaction of the plates was simulated by a strong attractive potential - SAT [25], given by

$$\frac{U_{SAT}}{\varepsilon} = \begin{cases} D_1\left[(\sigma/r_{ij})^{12} - (\sigma/r_{ij})^6\right] + D_2\,(r_{ij}/\sigma) - \varepsilon_s, & r_{ij} \leq r_{cs} \\ 0, & r_{ij} > r_{cs} \,, \end{cases} \tag{3}$$

where $r_{cs} = 2,0$ e $\varepsilon_s = D_1[(\sigma/r_{cs})^{12} - (\sigma/r_{cs})^6] + D_2(r_{cs}/\sigma)$. The parameters D_1 e D_2 are 1.2 and 0.545, respectively.

In the case of heterogeneous plates, around 20% random particles of the plates have hydrophobic wall-fluid interaction and the others, hydrophilic behavior.

The figure 3 illustrates the WCA and the SAT particle-plate interaction potentials.

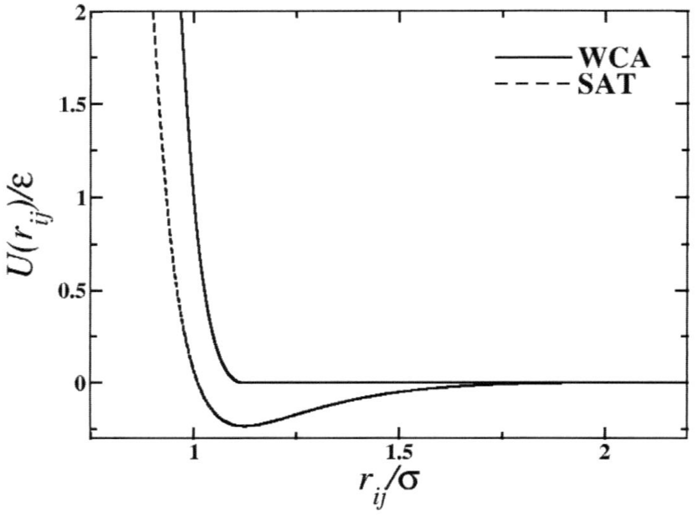

Figure 3. Particle-plate interaction potentials: hydrophobic, WCA (solid line), and hydrophilic, SAT (dashed-line).

3. The Methods and Simulation Details

The plates are composed by $N = 450$ spherical particles and they are immersed in the water-like fluid that is made by $N = 1700$ particles. The effective distance between the plates is fixed. Here the studied distances are $d = 3\sigma$, $d = 6.0\sigma$ and $d = 10\sigma$. The density of the systems was calculated considering just the water-like particles, excluding the particles of the wall. Due to the empty space near to each plate, the density needs to be corrected to an effective density, given by $\rho = N/(L^2 L_e)$, where $L_e = L - 2$ and L is the size of simulation box.

Molecular dynamics (MD) simulations at NVT-constant ensemble were performed to study the different systems. In order to keep fixed the temperature, the Nosé-Hoover thermostat [27, 28] with coupling parameter $Q = 2$ was used. The particle-particle interaction potential has a cutoff radius of $r_c = 3.5$ and it was shifted in order to have $U = 0$ at r_c.

Several densities and temperatures were simulated. Initial configuration in the solid and the liquid phases were selected and in both cases the equilibrium states were reached after 2×10^6 steps of simulation, followed by 4×10^6 simulation run. The time step was 0.001 in reduced unit. The average of all physical quantities were obtained every 50 time step just after the systems to be equilibrated. The metastability was avoided analyzing the pressure versus density phases diagram and the energy after the equilibrium state.

The structure of the systems was analyzed by the lateral radial distribution function, $g_{\parallel}(r_{\parallel})$, for each layer between the plates. An usual definition for $g_{\parallel}(r_{\parallel})$ is

$$g_{\parallel}(r_{\parallel}) \equiv \frac{1}{\rho^2 V} \sum_{i \neq j} \delta(r_{\parallel} - r_{ij}) \left[\theta\left(|z_i - z_j|\right) - \theta\left(|z_i - z_j| - \delta z\right)\right]. \tag{4}$$

where $r_{\parallel}$ is the parallel distance between the particles, considering just the x and y directions. The $\theta(x)$ is the Heaviside function and it restricts the sum of particle pairs in the same slab of thickness $\delta z = \sigma$.

The physical quantities in this paper are depict in LJ units [29],

$$r^* \equiv \frac{r}{\sigma}, \quad \rho^* \equiv \rho\sigma^3, \quad t^* \equiv t\left(\frac{\varepsilon}{m\sigma^2}\right)^{1/2} \quad \text{and} \quad T^* \equiv \frac{k_B T}{\varepsilon}, \tag{5}$$

for distance, density of particles, time and temperature, respectively.

Since we will use only reduced units in this paper, the $*$ will be omitted in the results discussion.

4. Results

The effects of the temperature and density on the fluid structure around the immersed parallel plates were investigated for the three types of plates. The following temperatures were analyzed: $T = 0.100$, 0.200 and 0.300. The densities studied correspond to sizes of simulation box in dimensionless units given by $L = 24$, 26 and 28, what correspond to the densities $\rho = 0.134$, 0.105 and 0.083, respectively also in dimensionless units.

4.1. Hydrophilic plates

In this subsection the distribution of the liquid under confinement with the hydrophilic plates is presented. The three densities analyzed show a very similar result and, for simplicity, we will show only results for $\rho = 0.134$. For this density, the transversal density profile is shown in the figure 4 for plates separated by (a) $d = 3$, (b) $d = 6.0$ and (c) $d = 10.0$. In all the cases layers are formed. We observed two, three and five layers for plates separated by (a) $d = 3$, (b) 6 and (c) 10, respectively.

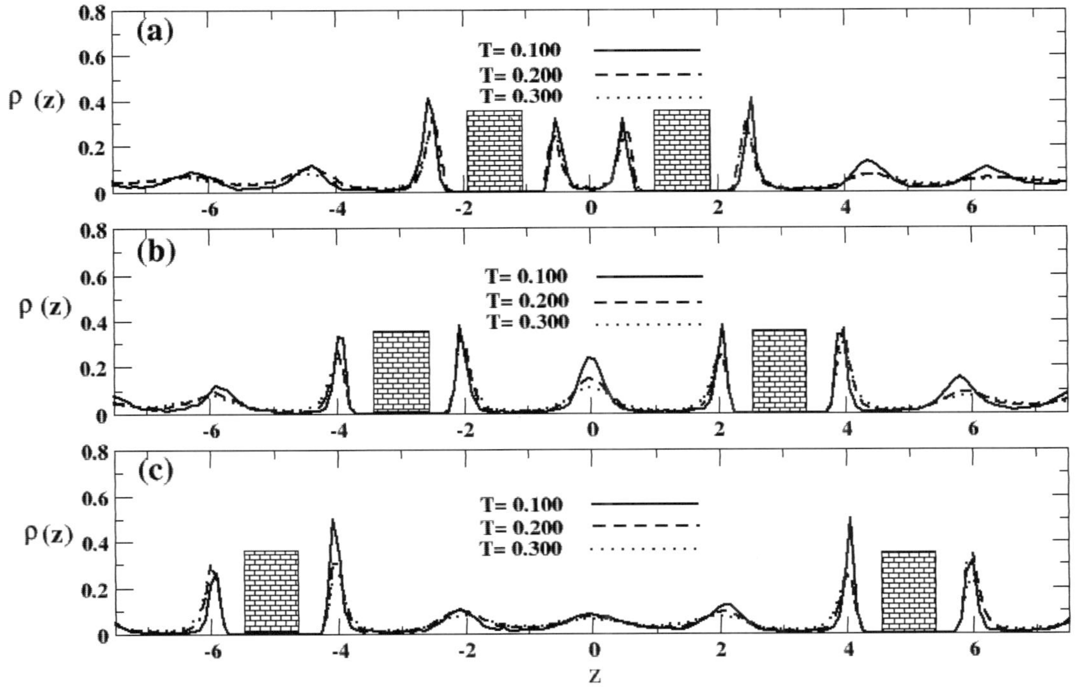

Figure 4. Tranversal density profiles for $\rho = 0.134$ and three different distances of plates: (a) $d = 3.0$, (b) $d = 6.0$ and (c) $d = 10.0$. Solid, dashed and dotted lines represent $T = 0.100$, $T = 0.200$ and $T = 0.300$, respectively.

In order to analyze the different structures of the layers the radial distribution functions,

$g_{||}(r_{||})$, for the contact layers were computed. The structure of the contact layers inside and outside the plates for $d = 3$, shown in the figure 5 (a) and (b), respectively, are very similar. Both layers are in a liquid-like phase. As the temperature is decreased, the liquid is more structured, what can be seen by the increasing of the first peak of $g_{||}(r_{||})$.

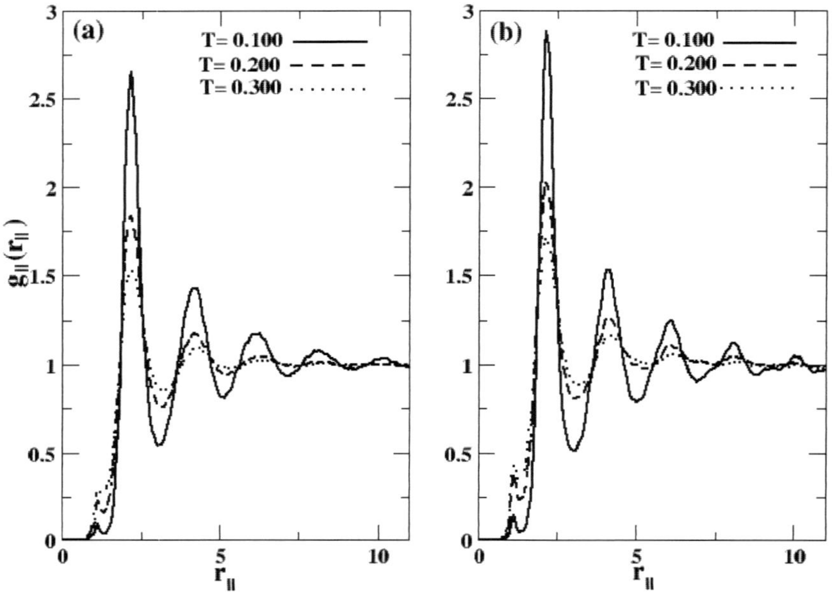

Figure 5. $g_{||}(r_{||})$ for contact layers (a) inside and (b) outside the hydrophilic plates separated by $d = 3$ for the indicated temperatures.

For the distance between the plates $d = 6$, the transversal density profile is shown in the figure 4 (b). In addition to the contact layers, there is the formation of a middle layer between the plates. As the temperature decreases, the density near the plates increases, what suggest that the fluid is more structured.

In figure 6 the radial distribution function $g_{||}(r_{||})$ for the contact and middle layers were computed. The middle layer for all the temperatures is liquid-like. The contact layer in region outside the plates is also liquid-like while the internal contact layer for low temperatures forms an amorphous phase, in agreement with previous results [23].

For plates separated by $d = 10$, the transversal density profiles are presented in figure 4 (c). The increases of density near the plates is similar to already observed for the other cases, $d = 3$ and $d = 6$. For these systems, all the layers present a liquid-like behavior and no amorphous phases was observed.

The presence of the amorphous phase at the contact layer for $d = 6$ and not for $d = 3$ and $d = 10$ is a consequence of the layering. The system with three layers exhibits a larger density at the contact layer leading to the formation of a solid-like phase.

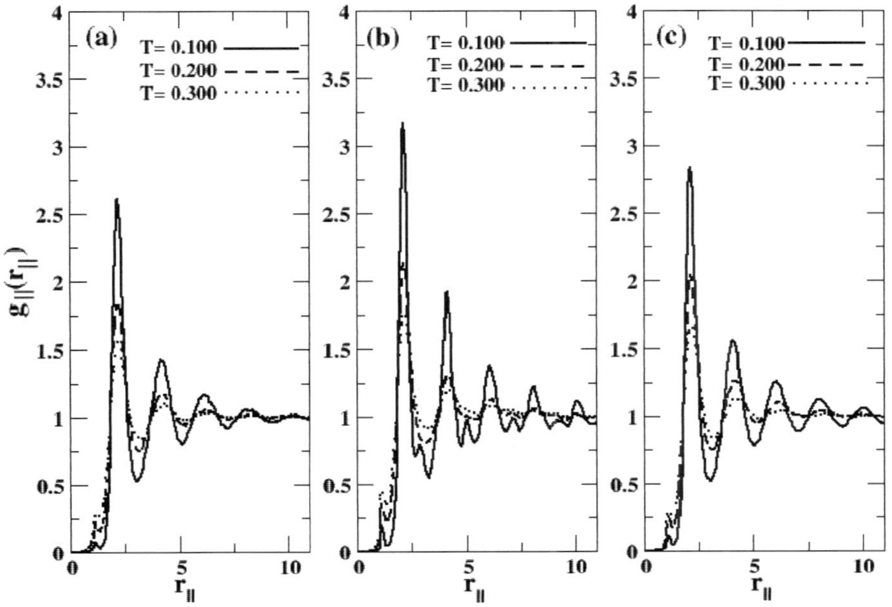

Figure 6. $g_{||}(r_{||})$ for a system with hydrophilic plates separated by $d = 6$ for (a) the contact layer outside of the plates, (b) the contact layer inside the plates and (c) the middle layer.

4.2. Hydrophobic plates

In this subsection the density profile of the liquid under confinement by the hydrophobic plates is shown. For simplicity, the analysis will be presented only for the density $\rho = 0.134$ and the distance between the plates $d = 6$. Figure 7 shows the layering structure observed for these systems, similar to observed for the hydrophilic confinement. It is possible to observe that for these systems, as the temperature decreases, the density of water-like particles increases near to each plate. These results are in agreement with observed for atomistic models of water confined in parallel plates [30] and hydrophobic surface of protein [31]. In order to investigate the structures of the particles of the layers, the radial distribution function, $g_{||}(r_{||})$, was computed for the contact and middle layers, as illustrated in the figure 8. The comparison between figure 8 and figure 6 shows differences between the fluid in the hydrophilic and hydrophobic confinement. While the hydrophilic confinement induces the formation of amorphous-like phases for low temperatures at $d = 6$, the hydrophobic system remain the fluid at low temperatures, what is in agreement with experimental results of Deschamps et al. [3] and Jelassi et al. [4].

4.3. Heterogeneous Parallel Plates

In the previous subsections, we studied the effects of the decrease in temperature in a homogeneous system of the fluid confined in hydrophobic or hydrophilic plates. However, in practical cases, immersed surfaces are expected to be chemically heterogeneous, that is, with hydrophilic and hydrophobic regions. First, we found a similar behavior to observed in the

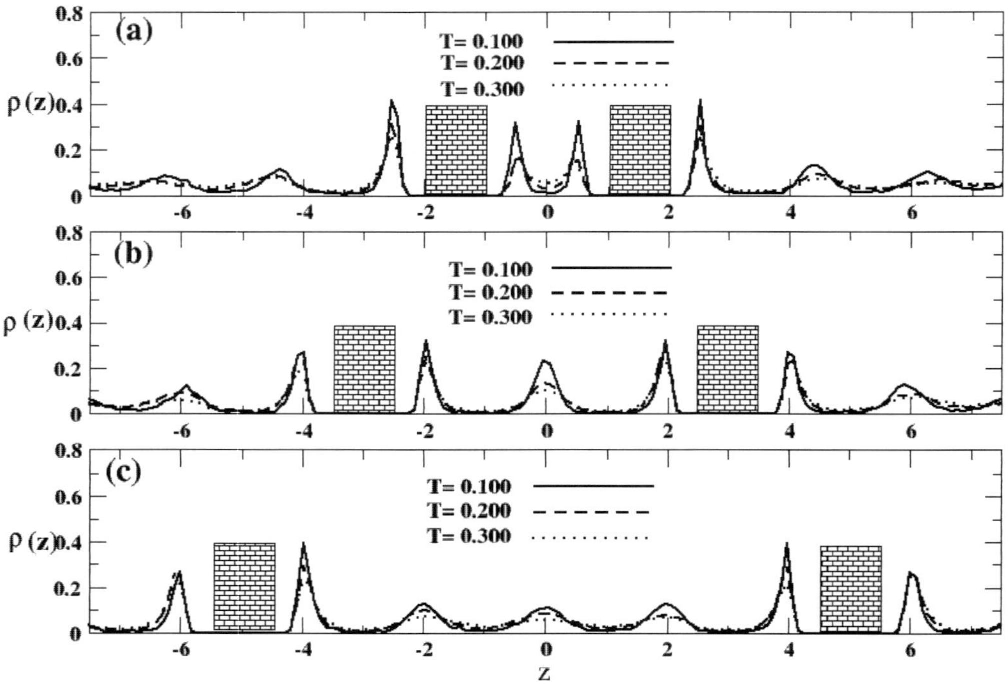

Figure 7. Density profiles for three different distances of hydrophobic plates. (a) $d = 3.0$, (b) $d = 6.0$ and (c) $d = 10.0$. Solid, dashed and dotted lines represent $T = 0.100$, 0.200 and 0.300, respectively.

homogeneous plates, that is, the density increases upon cooling also near to heterogeneous surface. We also studied the structural properties of the confined and not confined water-like particles.

The transversal density profile presents a similar behavior to other cases and we will not present for simplicity. The structure was analyzed using the $g_{||}(r_{||})$, shown in the figure 9, for $d = 6$ and $\rho = 0.134$. The other cases were not shown for simplicity. As the temperature decreases, the density of particles near to plates increases and the systems becomes more structured. Again, as observed in the hydrophilic homogeneous system, the contact layer inside the plates presents an amorphous-like behavior, while the other layers present a liquid-like behavior. It happens because the heterogeneous plates have hydrophilic interaction predominating over the hydrophobic one.

5. Conclusions

In this work we studied the effect of temperatures on the interfacial properties of water immersed in a parallel plates using molecular dynamics simulations. Hydrophilic, hydrophobic and heterogeneous surfaces were studied. The formation of four layers, two in confined part and two in the not confined part of the system are observed for small values of d. As we increase the distance between the plates we observe other layers besides of the contact layers. These layers we call middle layers. Our results suggest that layers are formed in order to minimize the

Journal of Physics: Conference Series **686** (2016) 012004 doi:10.1088/1742-6596/686/1/012004

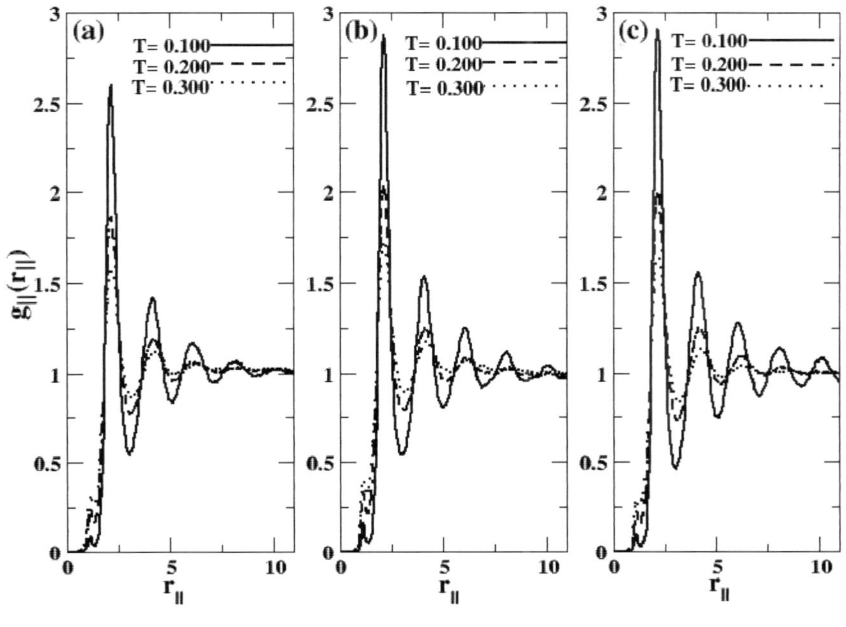

Figure 8. $g_{||}(r_{||})$ for water-like fluid in hydrophobic plates separated by $d = 6$, at $\rho = 0.134$, for (a) the contact layer outside of the plates, (b) the contact layer inside of the plates and (c) the middle layer.

particle-particle and the particle-plates interaction.

We also studied the structural properties of confined and not confined fluid. We calculate the lateral radial distribution function for each layer of the different regions of the system. The effects on the structure of water-like particles is better observed as the temperature decreases. The water-like particles in unconfined region has fluid characteristics in all cases analyzed. To the confined region we can observe that the contact layer to the hydrophilic plate has a liquid-like characteristic at $T = 0.300$ and $T = 0.200$, while at $T = 0.100$ the system presents an amorphous-like behavior. As the temperature is decreased the particles become more structured at the confined region. However, water-like particles confined by hydrophobic plates present a liquid-type characteristics.

In order to investigate the effects of plates type interaction with the fluid, we created a system with heterogeneous parallel plates, that is, plates with hydrophilic and hydrophobic interaction with fluid particles. These heterogeneous systems, that have around 80% of hydrophilic interaction, present results very similar to that observed for purely hydrophilic system. As the temperature decreases the system becomes more structured and at $T = 0.100$ the system presents an amorphous-type characteristic.

References

[1] Waller R 1964 *Essays of naturals experiments* (Nova Iorque)
[2] van Beest B, Kramer G and van Santen R 1990 *Phs. Rev. Lett.* **64** 1955

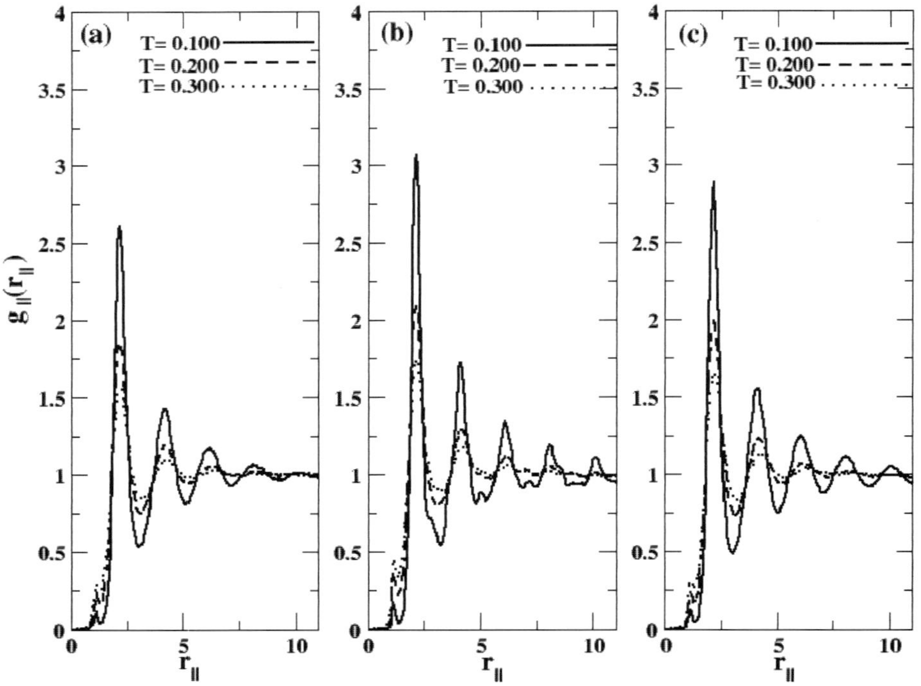

Figure 9. $g_{||}(r_{||})$ for water-like fluid in heterogeneous plates for $d = 6$ and $\rho = 0.134$. (a) is the contact layer outside of the plates, (b) is the contact layer inside the plates and (c) the middle layer.

[3] Deschamps J, Audonnet F, Brodie-Linder N, Schoeffel M and Alba-Simionesco C 2010 *Phys. Chem. Chem. Phys.* **12** 1440
[4] Jelassi J, Grosz T, Bako I, Belissent-Funel M C, Dore J C, Castricum H L and Sridi-Dorbez R 2011 *J. Chem. Phys.* **134** 064509
[5] Sliwinska-Bartkowiak M, Jazdzewska M, Huang L L and Gubbins K E 2008 *Phys. Chem. Chem. Phys.* **10** 4909
[6] Farimani A B and Aluru N R 2011 *J. Phys. Chem. B* **115** 12145
[7] Koga K, Gao G T, Tanaka H and Zeng X C 2001 *Nature (London)* **412** 802
[8] Bordin J R, Diehl A and Barbosa M C 2013 *J. Phys. Chem. B* **117** 7047–7056
[9] Giovambattista N, Rossky P J and Debenedetti P G 2009 *Phys. Rev. Lett.* **102** 050603
[10] Bordin J R, Krott L B and Barbosa M C 2014 *J. Phys. Chem. C* **118** 9497
[11] Bordin J R, Krott L B and Barbosa M C 2014 *J. Chem. Phys.* **141** 144502
[12] Strekalova E G, Mazza M G, Stanley H E and Franzese G 2012 *Journal of Physics: Condensed Matter* **24** 064111 URL `http://stacks.iop.org/0953-8984/24/i=6/a=064111`
[13] Morishige K and Iwasaki H 2003 *Langmuir* **19** 2808
[14] Morishige K and Kawano K 1999 *J. Chem. Phys.* **110** 4867
[15] Erko M, Findenegg G H, Cade N, Michette A G and Paris O 2011 *Phys. Rev. B* **84** 104205
[16] Xu L, Giovambattista N, Buldyrev S V, Debenedetti P G and Stanley H E 2011 *J. Chem. Phys.* **134** 064507
[17] de Oliveira A B, Netz P A, Colla T and Barbosa M C 2006 *J. Chem. Phys.* **124** 084505
[18] Buldyrev S V, Franzese G, Giovambattista N, Malescio G, Sadr-Lahijany M R, Scala A, Skibinsky A and Stanley H E 2002 *Physica A* **304** 23
[19] Barraz Jr N, Salcedo E and Barbosa M 2009 *J. Chem. Phys.* **131** 094504
[20] Leoni F and Franzese G 2014 *J. Chem. Phys.* **141** 174501

[21] Dudalov D E, Fomin Y D, Tsiok E N and Ryzhov V N 2014 *Phys. Rev. Lett.* **112** 157803
[22] Dudalov D E, Fomin Y D, Tsiok E N and Ryzhov V N 2014 *Soft Matter* **10** 4966
[23] Krott L B and Barbosa M C 2013 *J. Chem. Phys.* **138** 084505
[24] de Oliveira A B, Netz P A, Colla T and Barbosa M C 2006 *J. Chem. Phys.* **125** 124503
[25] Krott L B and Barbosa M C 2014 *Phys. Rev. E* **89** 012110
[26] Weeks J D, Chandler D and Andersen H C 1971 *J. Chem. Phys.* **54** 5237
[27] Hoover W G 1985 *Phys. Rev. A* **31** 1695
[28] Hoover W G 1986 *Phys. Rev. A* **34** 2499
[29] Allen M P and Tildesley D J 1987 *Computer Simulations of Liquids* 1st ed (Oxford: Claredon Press)
[30] N Giovambattista P J R and Debenedetti P G 2006 *Phys. Rev. E* **73** 041604
[31] Barbosa R and Barbosa M 2015 *Physica A* **439** 48

Langevin simulations of stripe forming systems with long-range isotropic competing interactions

Lucas Nicolao
Departamento de Física, Universidade Federal de Santa Catarina, 88040-970, Florianópolis, Brazil.
E-mail: lucas.nicolao@ufsc.br

Alejandro Mendoza-Coto
Departamento de Física, Universidade Federal do Rio Grande do Sul, Caixa Postal 15051, 91501-970, Porto Alegre, Brazil.

Daniel A. Stariolo
Departamento de Física, Universidade Federal Fluminense and National Institute of Science and Technology for Complex Systems, 24210-346, Niterói, Brazil.

Abstract.
We address the critical properties of the isotropic-nematic phase transition in stripe forming systems. We focus on isotropic models in which a short range attractive interaction competes with a long range repulsive interaction decaying as a power-law, paying particular attention to the cases in which this interaction is of dipolar or Coulomb types. Using Langevin dynamics simulations we show that these models belong to different universality classes. The numerical algorithms developed for each case are presented and discussed in detail. The obtained results are in agreement with recent theoretical predictions [1], according to which the orientational transition belongs to the Kosterlitz-Thouless universality class for dipolar systems, while it is of the second-order type for Coulomb systems.

1. Introduction

Competing interactions usually lead to heterogeneities in the local order parameter in the form of completely disordered patterns as present, e.g. in spin glasses [2], or formation of patterns or domains as observed in many systems like magnetic thin films [3], diblock copolymers [4], and quasi two dimensional electronic quantum liquids [5]. The identification and characterisation of ordered phases in these systems is not straightforward because of the spatially heterogeneous nature of the order parameter (see Figure 1 for an experimental example). Broken translation and rotation symmetry in space are usually associated with phase transitions. Early theoretical descriptions of phase transitions in these systems exploited an analogy with liquid crystals, systems of more or less rigid rod-like molecules which can order in a large variety of phases with broken translation or rotation symmetry. Like in liquid crystals, where orientation of the axes

Journal of Physics: Conference Series **686** (2016) 012005 doi:10.1088/1742-6596/686/1/012005

of the molecules are described by a vector called "director", in pattern forming systems like ultrathin magnetic films a director can be associated to the direction of interfaces or domain walls and are useful to identify orientational order of the patterns (see Figure 2).

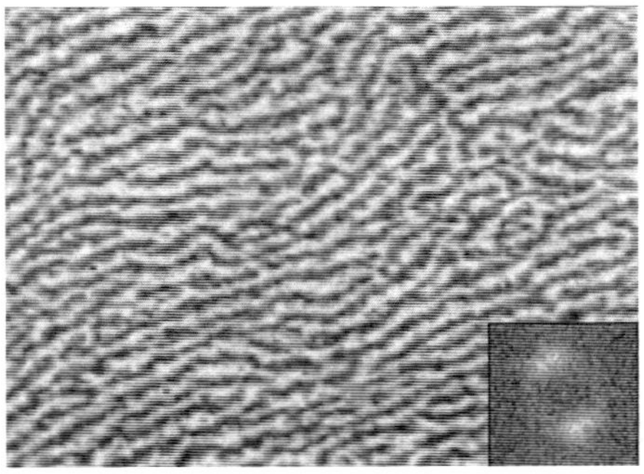

Figure 1. Stripe domains of magnetization in ultrathin magnetic films of 2.6 ± 0.6 AL thick fcc Fe on Cu(100), as imaged by scanning electron microscopy with polarisation analysis (SEMPA). This image's horizontal dimension is $\simeq 24\ \mu$m. Inset: Fourier power spectrum. Adapted from [6].

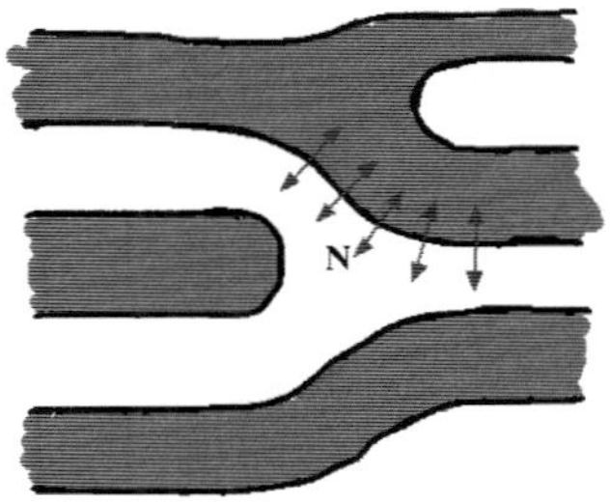

Figure 2. Schematic representation of a striped pattern with local orientation of interfaces characterised by the director vector.

In a pioneering work in the seventies S. A. Brazovskii showed that some antiferromagnets with nearly isotropic competing interactions acting at different scales suffer an instability leading to anisotropic low temperature phases with modulation of the local order parameter in the form of charge or spin density waves [7]. The instability occurs at a characteristic non zero wave vector k_0 leading precisely to modulated structures in the charge density or magnetization. This kind of instability was realised to be of much more general applicability and soon after Brazovskii's work J. Swift and P. C. Hohenberg proposed a similar mechanism to explain the convective instability manifested by the appearance of rolls in the Rayleigh-Bénard convection [8]. Motivated by the recent discovery of high temperature superconductivity, V. J. Emery and S. A. Kivelson introduced in the early nineties a model with competition between a kinetic term and the long-range Coulomb interaction between electrons which leads also to phase separation in the charge density and the appearance of modulated order in the form of stripes[9]. In the following years experiments on high T_c superconductors, doped Mott insulators and quantum Hall systems established the presence of anisotropies in transport properties, and new states of matter at ultra low temperatures, the so called "electronic liquid crystals", were proposed to explain the observations [10, 11, 12]. Since then, the presence of stripe, nematic or hexatic order in the charge and/or spin degrees of freedom of quasi two dimensional strongly correlated electronic systems has been firmly established as part of the phenomenology around high T_c superconductivity, although its relevance to the appearance of superconducting properties is still a matter of debate [13, 14].

Despite the large amount of experimental and theoretical work done on different systems showing modulated order, still there are many properties of the phase diagrams and phase transitions which are not known. In particular, the characterisation of the nature of different phases and the critical properties have been shown to be subtle. Experiments able to probe the critical properties of phase transitions in these systems have still to be designed and executed, while theoretical approaches suffer from some well known but severe limitations in their predictability. First of all, the nearly two dimensional nature of the systems makes the precise characterisation of their critical properties a subtle task. It is known that two dimensions is near the lower critical dimension of many systems and the fluctuation effects are rather strong [15], making predictions from naive mean field approximations or simulations of moderately small system sizes unreliable. Furthermore, the very nature of the *nematic phases*, phases with orientational but not translational order, leads to an order parameter which measures anisotropies in nearest-neighbour correlation functions, i.e. the order parameter is a two-point function in space, and then simple mean field theories are unable to capture the presence of a nematic phase.

Another important issue that hinders the full description of phase diagrams is the fact that the relevant scales can vary many orders of magnitude by tuning an external parameter and that is true even far from critical points. In general the modulation length of stripes or orientationally ordered states depends on external parameters like temperature and magnetic fields, but mainly on the relative strength of the competing interactions. In this way, the modulation length can range from a few interatomic or lattice spacings up to hundreds or thousands of them. Examples of systems in which the relevant physics corresponds to one or the other extreme can be found [16]. Clearly, the critical properties in both cases can be very different, anisotropies induced by the lattice structure are relevant when the modulation length is of the order of the lattice spacing, but probably not relevant for modulations spanning thousands of lattice spacings. This multiscale phenomenology strongly complicates the modelling and simulation of stripe forming systems.

Here we address the critical properties of stripe forming systems with isotropic long-range competing interactions with variable range [1]. We give special attention to the high temperature phase transition between an isotropic and nematic phases, and the relevance of the range of the interaction to the nature of this rotational symmetry breaking transition. In the next section we motivate our general mesoscopic approach to the two pertinent stripe forming systems, with Coulomb or dipolar interactions, discussing general issues about the nematic phases in these systems. We also review a recently developed elastic theory that accounts for the correct effect of long-range interactions, predicting distinct behaviour for the two cases. Later we present the technical details involved in performing Langevin simulations of these models. We finish presenting preliminary results confirming these predictions, and discussing future directions.

2. Coarse-grained model and elastic theory

The long wavelength limit, in which the modulation spans thousands of electrons or spins, is particularly relevant in two well known physical systems: ultrathin magnetic films [6, 17, 18] and two dimensional electronic liquids [10, 19, 20, 21]. In the first case, stripes or orientated patterns emerge from the competition between the short range exchange interaction and the long-range dipolar interaction. In the second case, the nature of the competition is more subtle coming from the competition between the kinetic (quantum dynamics) term and the long-range Coulomb interaction. In this limit, both cases can be modelled by an effective coarse-grained Hamiltonian in two dimensions of the form:

$$\mathcal{H}[\phi(\vec{x})] = \frac{1}{2}\int d^2x \left(\vec{\nabla}\phi(\vec{x})\right)^2 + \frac{1}{2}\int d^2x \int d^2x'\ \phi(\vec{x})J(\vec{x}-\vec{x'})\phi(\vec{x'}) + \frac{1}{2\beta}\int d^2x\ V(\phi(\vec{x})) \quad (1)$$

where $\beta = 1/k_B T$ and $V(\phi) = -\frac{r}{2}\phi^2 + \frac{1}{12}\phi^4$ is a local potential. The first term can be seen as the continuous limit of a nearest-neighbour interaction while the second, competing term, has the form $J(\vec{x}) = J/x^\alpha$, which allows to analyse in a unified way short range (large α) and long-range (small α) interactions. Physically relevant examples are the Coulomb interaction

($\alpha = 1$) and the dipolar interaction between out of plane magnetic moments ($\alpha = 3$). Note that this effective Hamiltonian has strictly isotropic interactions and the anisotropic phases, like the nematic one, can only emerge from the spontaneous breaking of the rotational symmetry in real space. In fact, working with a similar Hamiltonian, D. G. Barci and D. A. Stariolo [22] showed that the rotational symmetry of the characteristic wave vector $\vec{k}_0$ present at relatively high temperatures is spontaneously broken at a critical temperature T_c leading to a isotropic-nematic phase transition. Nematic order in this case means that the stripes possess an orientational order

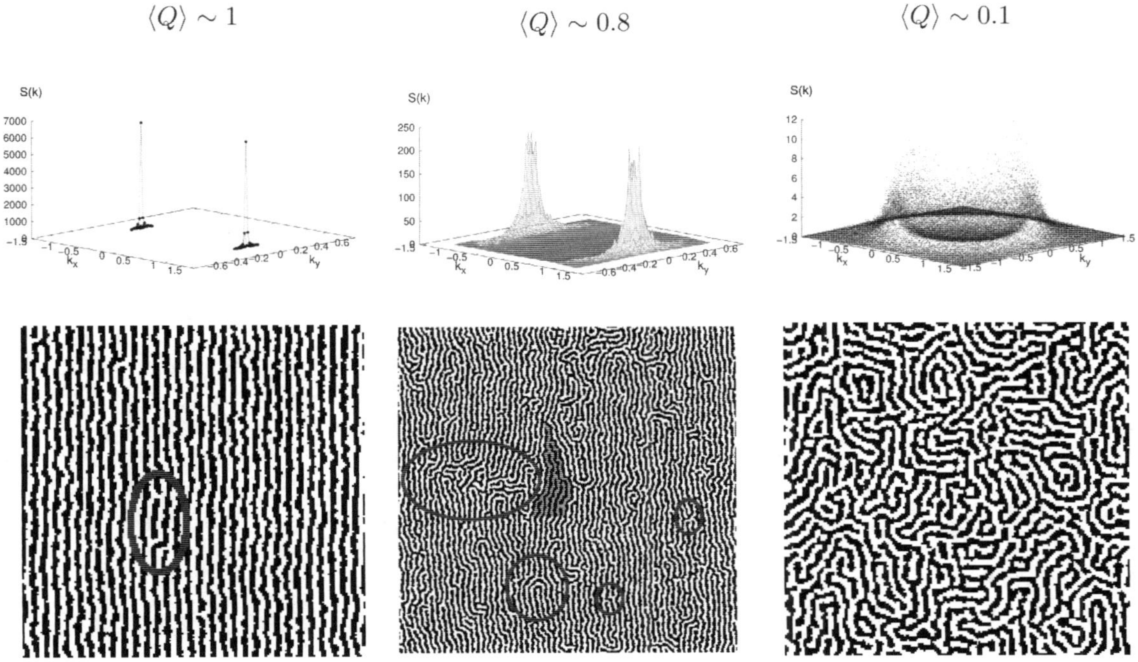

Figure 3. Results from Langevin simulations of the dipolar model at three temperatures, increasing from left to right. The stripe width span three lattice sites. Here are shown some representative configurations with sizes 192^2 (384^2 middle) with some topological defects highlighted, structure factors and mean orientational order parameter values corresponding to smectic, nematic and isotropic phases, respectively. Taken from [23].

but there is no periodicity under spatial translations. The nature of the nematic-isotropic phase transition was predicted to be of the Kosterlitz-Thouless type based on physically motivated arguments [24] and Langevin simulations on the dipolar case [23] gave support on this finding, in spite of some limitations due to numerical lattice induced anisotropies. Some results from this Langevin simulations, shown in Figure 3, are particularly relevant to illustrate the nematic order and its associated order parameter and angle fluctuations in the structure factor.

The recent introduction of a new elastic theory allowed us to identify the degrees of freedom at the relevant (wavelength) scale and to make precise analytical predictions on the nature of phase transitions in stripe forming systems in two dimensions, for arbitrary values of the range of the competing interaction $1/r^\alpha$ [1]. Allowing for the phase of the modulation to fluctuate in a way that the local order parameter is given by $\phi(\vec{x}) = \sum_n \phi_n \cos(nk_0 x + nk_0 u(\vec{x}))$, the effective Hamiltonian for the deformation field $u(\vec{x})$ has the following form:

$$\mathcal{H}_p = \frac{1}{2} \int \frac{d^2k}{(2\pi)^2} (\gamma_x k_x^2 + \gamma_y k_y^4 + \gamma_{nl} k^{\alpha-2} {k_y}^4) \hat{u}(\vec{k}) \hat{u}(-\vec{k}). \tag{2}$$

The first two contributions of this effective Hamiltonian correspond to the well known local elastic response of the stripe pattern [25], while the last contribution is related with the residual interaction between the stripes due to the presence of a long range repulsive interaction of the form $1/r^{\alpha}$ [1]. It is possible to verify that for repulsive interactions sufficiently long-ranged ($\alpha < 2$), the non-local contribution in the effective Hamiltonian (2) dominates over the transversal local contribution. This is an unexplored regime in the literature, since previous works always assumed that even in the presence of long-range interactions the effective Hamiltonian possess only local contributions.

Nevertheless, a careful analysis of the positional fluctuations at a Gaussian level reveals that $\langle u^2 \rangle$ diverges linearly with the linear dimension of the system for $\sigma \geq 2$ and with a power $\frac{\sigma}{2}$ for $\sigma < 2$. This means that, independently of the range of the interaction, positional order is unstable.

On the other hand, it was proven that at scales much larger than the modulation length the effective orientational Hamiltonian in terms of the normalised local average of the director field $\vec{n}(\vec{x})$ is given by:

$$\begin{aligned}\mathcal{H}_o &= \frac{\gamma_1}{2}\int d^2x\,(\partial_\mu \vec{n}(\vec{x}))^2 + \frac{\gamma}{2}\int d^2x \int d^2x'\Omega(|\vec{x}-\vec{x}'|)\left(\frac{\vec{n}(\vec{x})\cdot\vec{n}(\vec{x}')}{|\vec{x}-\vec{x}'|^{\sigma+2}}\right.\\ &- \left.(\sigma+2)(\frac{\vec{n}(\vec{x})\cdot(\vec{x}-\vec{x}')\vec{n}(\vec{x}')\cdot(\vec{x}-\vec{x}')}{|\vec{x}-\vec{x}'|^{\sigma+4}}\right),\end{aligned} \tag{3}$$

where γ_1 and γ represent the local and the non-local stiffnesses, respectively, and $\Omega(x)$ represents a short distance cut-off function. This Hamiltonian has been previously studied by Maier et.al. [26] for the case $\sigma = 1$. As pointed out by their results, the anisotropic term of the non-local dipolar interaction is irrelevant for the critical properties in two dimensions. This simplifies greatly the study to be made, since it makes it possible to replace the original non-local interaction by a ferromagnetic one decaying with the same power-law of that in the original orientational Hamiltonian without changing the critical properties of the model.

With this considerations in mind, it is straightforward to conclude that in systems with sufficiently long-ranged interactions ($\sigma < 2$) the non-local attractive interaction rules over the local contribution, stabilising a long-range orientational order at low temperatures. On the other hand, when $\sigma \geq 2$, the short ranged square gradient interaction dominates over the non-local contribution, resulting in the known Kosterlitz-Thouless (KT) transition scenario of the classical XY model [25].

For the new interesting case with $\sigma < 2$, the critical properties have been worked out by non-perturbative renormalization group techniques, and can be summarised as:

(i) in the critical region the correlation length diverges exponentially at T_c from both sides, as a reminiscent of the KT transition behaviour:

$$\xi_o \propto \exp\left(\frac{b}{\sqrt{|T_c - T|}}\right), \tag{4}$$

(ii) For $T < T_c$ in the critical region, the average order parameter behaves as $M \propto \xi_o(T)^{-(2-\alpha)/2}$, showing the existence of long-range order when $\alpha < 2$.

(iii) The orientational susceptibility diverges as $\chi_o \propto \xi_o(T)^{\alpha}$ in the critical region.

In the following section we present some of the main difficulties in exploring these and other features of two dimensional rotational symmetry breaking systems in numerical simulations. After a brief review of previous results about the nematic phase in the dipolar case, we present our method and recent results confirming the above scenario.

3. Simulations

In order to explore further this rich phenomenology in the complementary perspective of statistical physics simulations, it is important to have the proper tools for the phenomenon of interest. A traditional approach is to perform Monte Carlo simulations of the corresponding Ising model: $H = -\delta \sum_{<i,j>} s_i s_j + \sum_{(i,j)} s_i s_j / r_{ij}^{\alpha}$ [27, 28]. Even though this methodology is very useful to explore many interesting and general thermodynamic [29] and dynamical [30, 31] phenomena present in these stripe forming systems, their "microscopic" approach is affected by strong lattice effects. This lattice anisotropy is the result of the sharp (Ising) domain walls between each stripe in combination with the relative small stripe width (typically from 2 to 4 spins per stripe). In this conditions the system is dominated by the square lattice discrete symmetry, such that at high temperatures the disordered stripes form what is commonly called a tetragonal liquid [32]. The strong lattice induced anisotropy present in such simulations naturally prevent the study of rotational symmetry breaking phenomena of a priori isotropic models.

Instead, we take as starting point the same mesoscopic coarse-grained model (1), that can be seen as the effective, long wavelength limit of the corresponding Ising models. Simulations of field theory models such as these have been made extensively in the past, with Monte Carlo [33, 34], Langevin dynamics [35], or a combination of the two [36]. We have built on the methodology developed for the study of phase-ordering kinetics [37, 38, 39] and general dynamical critical phenomena [40]. Besides the many interesting dynamic features of the out-of-equilibrium dynamics of the stripe forming systems - a comprehensive review can be found in [41] - the same machinery can be applied to study their thermodynamic equilibrium properties.

Early studies of the nematic phase from Langevin simulations of the dipolar model revealed important large-scale features of a nematic phase, like anisotropic domains (shorter in the direction parallel to the stripes) and angle fluctuations (see the structure factor in Figure 3). On the other hand we observed long-range correlations in the nematic order instead of the expected quasi-long range order: fourfold lattice anisotropies, though softened because of the continuous nature of the order parameter field, were induced mainly by the short stripe width, of only three lattice sites per stripe. The effect of the four-fold anisotropy was to disrupt the orientational Goldstone excitation and changing the associated KT transition into a second-order phase transition [42, 43]. As it will become evident below, wider stripes are much less susceptible to such anisotropies at moderate temperatures. We now proceed to present the technical details of the Langevin simulations developed for this work.

3.1. Method

We performed Langevin simulations of the Hamiltonian (1). The relaxational (overdamped) Langevin dynamics of the density $\phi(\vec{x})$ is defined by:

$$\frac{\partial \phi(\vec{x})}{\partial t} = -\frac{\delta H[\phi]}{\delta \phi(\vec{x})} + \eta(\vec{x}, t), \tag{5}$$

which for the Hamiltonian in (1) reads:

$$\frac{\partial \phi}{\partial t}(\vec{x}, t) = \nabla^2 \phi - \frac{1}{\delta} \int d^2x' \frac{1}{|\vec{x} - \vec{x}'|^{\alpha}} \phi(\vec{x}', t) + r\phi(\vec{x}, t) - u\phi(\vec{x}, t)^3 + \eta(\vec{x}, t) \tag{6}$$

where δ and u are fixed parameters, $r = (T_c - T)$ and $\eta(\vec{x})$ represents a Gaussian white noise with correlations $\langle \eta(\vec{x}, t)\eta(\vec{x}', t') \rangle = 2T\delta(\vec{x} - \vec{x}')\delta(t - t')$, where T is the effective temperature of the heat bath. In the Fourier space, this equation reads:

$$\frac{\partial \phi}{\partial t}(\vec{k}, t) = -\mathcal{A}_\alpha(k)\phi(\vec{k}, t) - u\left[\phi^3\right]_F(\vec{k}, t) + \eta(\vec{k}, t) \tag{7}$$

where $\left[\phi^3\right]_F(\vec{k}, t)$ stands for the Fourier transform of $\phi^3(\vec{x}, t)$, and $\mathcal{A}_\alpha(k)$ encompasses both the short k^2 and long-range $\mathcal{J}_\alpha(k)$ interactions. We worked with two forms of this kernel, the first

$$\mathcal{A}_3(k) = a_2(k - k_0)^2 - r \tag{8}$$

corresponds to the functional form of a dipolar interaction, basically from the linear dependence[44] in k. The second form is

$$\mathcal{A}_1(k) = a_2(k^2 + 2k_0^3/k - 3k_0^2) - r, \tag{9}$$

corresponding to the Coulomb model $1/k$. The parameters are such that both models have the same value of $\mathcal{A}_\alpha(k)$ and curvature close to the minimum k_0. To ensure this, we have set $a_2 = 1$ for the dipolar and $a_2 = 1/3$ for the Coulomb model. In both cases $T_c = 1$ and $k_0 = 1$.

The general solution of equation (7) is:

$$\phi(t + dt, \vec{k}) = \phi(t, \vec{k})\, \mathrm{e}^{-\mathcal{A}_\alpha(k)\, dt} + \int_t^{t+dt} \left[-u\left[\phi^3\right]_F(\vec{k}, t') + \eta(\vec{k}, t')\right] \mathrm{e}^{-\mathcal{A}_\alpha(k)\,(t+dt-t')} dt' \tag{10}$$

The general assumption towards the numerical integration of this equation is to consider the nonlinear part to be nearly constant during the time-step, between t and $t + dt$. The same cannot be assumed for the noise. By the central limit theorem, this integral is another Gaussian white noise with zero mean and variance such that, the general solution can be written as:

$$\phi(t + dt, \vec{k}) \simeq \phi(t, \vec{k})\, \mathrm{e}^{-\mathcal{A}_\alpha(k)\, dt} - u\left[\frac{1 - \mathrm{e}^{-\mathcal{A}_\alpha(k)\, dt}}{\mathcal{A}_\alpha(k)}\right]\left[\phi^3\right]_F(\vec{k}, t) + \sqrt{\frac{1 - \mathrm{e}^{-2\mathcal{A}_\alpha(k)\, dt}}{2\mathcal{A}_\alpha(k)}}\, \mu(\vec{k}) \tag{11}$$

where $\mu(\vec{k})$ is the same Gaussian white noise integrated in time: $\langle \mu(\vec{x})\mu(\vec{x}')\rangle = 2T\delta(\vec{x} - \vec{x}')$. Since we are dealing with a system with many degrees of freedom, we always seek for simple methods of order one to minimize the computational cost.

3.2. Dipolar case

The Euler scheme can be obtained expanding to first order in dt so that:

$$\phi(t + dt, \vec{k}) \simeq \phi(t, \vec{k})\,[1 - \mathcal{A}_\alpha(k)\, dt] - dt\, u\left[\phi^3\right]_F(\vec{k}, t) + \sqrt{dt}\, \mu(\vec{k}) \tag{12}$$

where it was taken into account the same approximation to the variance of the noise. The stability of this method is very poor [45]. A typical treatment to enhance the stability is to implement it as a semi-implicit scheme, where the laplacian term from (8) is computed in $t+dt$:

$$(1 + dt\, k^2)\phi(t + dt, \vec{k}) = \phi(t, \vec{k})\,[1 + r\, dt - dt\, 2k - dt] + \left[-dt\, u\, \phi^3(\vec{x}, t) + \sqrt{dt}\, \mu(\vec{x})\right]_F(\vec{k}, t) \tag{13}$$

where we have taken $k_0 = a_2 = 1$ as mentioned before, and have explicitly written the Fourier transform of the white noise. This is very close to the final implementation, we only have to address the spatial discretization. In the adimensional form where $k/k_0 \to k$ [41], the periodicity of the stripes are set by the lattice constant dx of the 2D square grid with linear size $L = MN$, so that $\vec{k} = (k_x, k_y)$ with $k_i = 2\pi n_i/(dx\, L)$, $n_i = -L/2, \ldots, L/2$ and $dx = \pi/M$. Within this scheme, the stripe length span M lattice sites and the linear system size is such that contains N stripes. Then, full lattice update is accomplished in the Fourier space, with the aid of fast Fourier libraries [46] through:

$$\phi(t+dt,\vec{k}) = \frac{1 + r\,dt - dt\,2k - dt}{1 + dt\,k^2}\phi(t,\vec{k}) + \frac{1}{1+dt\,k^2}\left[-dt\,u\,\phi^3(\vec{x},t) + \sqrt{\frac{2T dt}{dx^2}}\,\tilde{\mu}\right]_F (\vec{k},t) \quad (14)$$

where $\tilde{\mu}$ is a Gaussian white noise with zero mean and unitary variance (in the discretization process Dirac deltas turn into Kronecker deltas: $\delta(\vec{x}-\vec{x}') \to \delta_{\vec{x},\vec{x}'}/dx^2$). Note that this scheme can be written in the form:

$$\phi(t+dt,\vec{k}) = \mathcal{L}(k)\phi(t,\vec{k}) + \mathcal{I}(k)\psi(t,\vec{k}) \quad (15)$$

where the real coeficients can be calculated at the start of the program $\mathcal{L}(k) \equiv (1 + r\,dt - dt\,2k - dt)/(1 + dt\,k^2)$, $\mathcal{I}(k) \equiv 1/(1 + dt\,k^2)$ and $\psi(t,\vec{k}) \equiv \left[-dt\,u\,\phi^3(\vec{x},t) + \sqrt{2Tdt/dx^2}\,\tilde{\mu}\right]_F$. This way the computational performance is dominated by the two forward FFT, L^2 updates and then one backward FFT. The main advantages of updating in the Fourier space are the decoupling of the equations (memory speedup) as well as the use of an isotropic form for the laplacian, minimizing one possible source of undesired lattice effects.

3.3. Coulomb case

One can readily see that, in the Coulomb case where the fluctuation spectrum (9) has one term inversely proportional to k, we would have $\phi(t+dt,\vec{k}) \propto \phi(t,\vec{k})/k$ and there would be an instability close to $k \sim 0$ that cannot be attenuated by any implicit term in an Euler scheme. Instead, we take advantage of the general solution given by (10) where $\phi(t+dt,\vec{k}) \propto \phi(t,\vec{k})\exp(-1/k)$ and this instability can be exponentially attenuated. In order to write the equation (11) in a simple form as in equation (15), we can see that we have know two different coefficients for the non-linear part:

$$\mathcal{I}_1(k) \equiv \left[\frac{1 - \mathrm{e}^{-\mathcal{A}_1(k)\,dt}}{\mathcal{A}_1(k)}\right] \quad \text{and} \quad \mathcal{I}_2(k) \equiv \sqrt{\frac{1 - \mathrm{e}^{-2\mathcal{A}_1(k)\,dt}}{2\mathcal{A}_1(k)}} \quad (16)$$

for the contributions of the ϕ^3 and the μ terms, respectively. We can notice that both this terms are similar for small dt, especially if we write them in power series: $\mathcal{I}_1(k) \simeq dt - \mathcal{A}_1(k)dt^2/2 + \mathcal{A}_1^2(k)dt^3/6 + \ldots$, and $\mathcal{I}_2(k) \simeq dt^{1/2} - \mathcal{A}_1(k)dt^{3/2}/2 + \mathcal{A}_1^2(k)dt^{5/2}/4.8 + \ldots$, so that up to second order in dt we can assume $\mathcal{I}_2(k) \simeq \mathcal{I}_1(k)/\sqrt{dt}$. Using this approximation for the coefficient of the thermal noise we have the same update form as in equation (15), with:

$$\mathcal{L}(k) \equiv \mathrm{e}^{-\mathcal{A}_1(k)\,dt} \quad (17)$$

$$\mathcal{I}(k) \equiv \frac{1 - \mathrm{e}^{-\mathcal{A}_1(k)\,dt}}{\mathcal{A}_1(k)} \quad (18)$$

$$\psi(t,\vec{k}) \equiv \left[-dt\,u\,\phi^3(\vec{x},t) + \sqrt{\frac{2T}{dt\,dx^2}}\,\tilde{\mu}\right]_F \quad (19)$$

3.4. Simulation results

With suitable stable schemes for the integration of the equations, the next step it to ensure an suitable space discretization to explore the properties of the nematic phases. By exploring the temperature range with cooling protocols with small system size, we could estimate the liquid-nematic phase transition. After an estimation of the equilibration and correlation times from high temperature quench experiments, we could safely guarantee that below $T \simeq 0.59$ ($T \simeq 0.44$) the configurations find themselves in the low temperature phase (with orientational order) for all system sizes for the dipolar (Coulomb) model. Within this low temperature phase but still in the neighbourhood of the phase transition, we have measured the structure factor for various thermal histories and different stripe widths M, related to space discretization by $dx = \pi/M$, from $M = 3$ up to $M = 15$. We have obtained an isotropic average structure factor for $M > 7$, meaning the stripes don't have a preferred direction in the nematic phase, so that lattice anisotropies are irrelevant. A conservative choice along this work was $M = 11$ in order to assure smooth domain walls.

Since above the aforementioned temperatures the configurations are in the stripe-liquid phase, where both positional and orientational correlation lengths are finite, we have concentrated our efforts on equilibrium simulations inside the nematic phase, with $T = 0.57$ ($T = 0.43$) for the dipolar (Coulomb) model, for system sizes ranging from $(12 \times 11)^2$ up to $(66 \times 11)^2$.

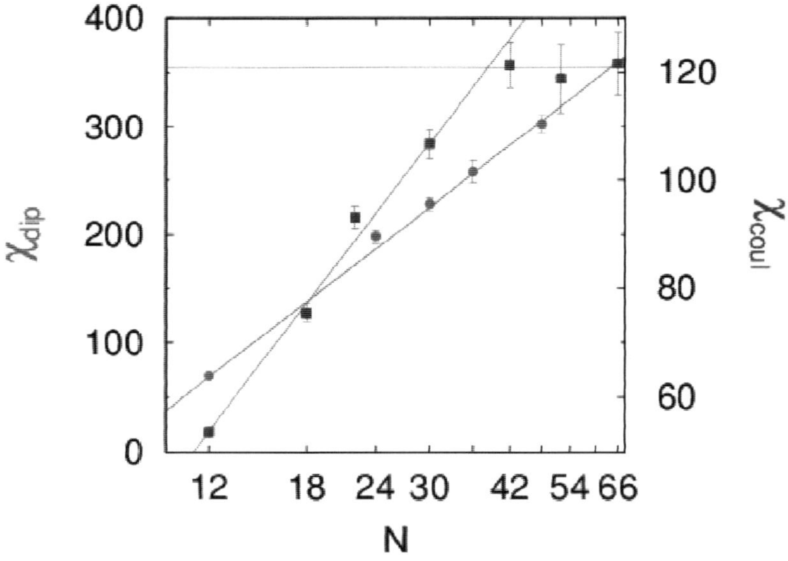

Figure 4. Orientational susceptibility in linear-log scale plot of the Coulomb (right) and dipolar (left) models as function of the linear size of the systems (normalized by half of the modulation length N=L/11). The full lines are fits to logarithmic function $A + B \log N$, with: $A = -355(18)$ and $B = 170(6)$ ($\chi^2/N_{dof} = 1.0$) in the dipolar case; $A = -362(10)$ and $B = 173(4)$ ($\chi^2/N_{dof} = 2.3$) in the Coulomb case. In both cases the number of degrees of freedom is $N_{dof} = 3$. The horizontal line is only a guide to the eyes.

In order to quantify the orientational order, through the local director field $\vec{n}(\vec{x}) =$

$\vec{\nabla}\phi(\vec{x})/|\vec{\nabla}\phi(\vec{x})|$ we have measured the nematic tensor:

$$Q_{\alpha\beta} = \sum_{\vec{x}} \left(n_a(\vec{x}) n_a(\vec{x}) - \frac{1}{2} \delta_{\alpha\beta} \right) \tag{20}$$

where $\alpha, \beta = 1, 2$ are the Cartesian components. The orientational order parameter is given by $Q = \sqrt{Q_{11}^2 + Q_{12}^2}$. We have measured this quantity and its corresponding susceptibility χ using between 5 and 25 thermal histories, resulting approximately in an overall $2 \times 10^6/N^2$ ($3 \times 10^6/N^2$) independent measurements for the dipolar (Coulomb) model - recalling that N is the number of stripes that can fit in the linear size of the system, both related by $N = L/11$, so the number of samples lies roughly between 700 and 20000.

Our results are shown in Figures 4 and 5. We can readily see from Figure 4 that below the transition temperature the susceptibility of the corresponding order parameter presents a limited growth up to a plateau for the Coulomb case. This is what is expected in a second order phase transition. On the other hand, with dipolar interactions we observe a monotonic (probably logarithmic) increase of the susceptibility with the system size, without any sign of attenuation. This is in accordance with a Kosterlitz-Thouless phase, where the susceptibility should diverge in the thermodynamic limit for all $T \leq T_{KT}$. Even though we show logarithmic fits along the points in Figure 4, the small number of points don't allow to clearly distinguish this functionality from power-laws, algebraic fits are equally good [1] specially for the Coulomb case.

Of course, this is an analysis for a single low temperature and it presents a limited power if compared to more traditional finite-size scaling analysis. Indeed, we cannot conclude that the dipolar χ will not saturate at larger N's, but the different trend observed in both systems for equivalent parameter values is a strong indication that the theoretical results are indeed correct. In Figure 5 the value of the orientational order parameter as function of the linear size for both models are similar in behaviour, showing how subtle the characterization of both models can be.

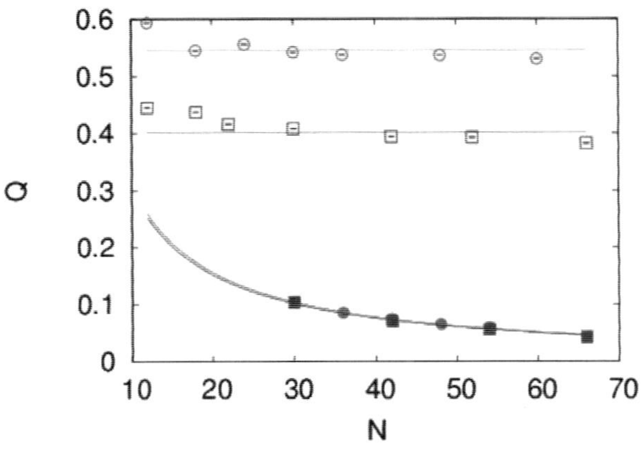

Figure 5. Orientational order parameter dependence on the normalized linear size of the systems for the Coulomb (blue) and dipolar (red) models, for temperatures above (filled) and below Tc (open symbols). These temperatures correspond to $T = 0.60$ ($T = 0.45$) and $T = 0.57$ ($T = 0.43$) for the dipolar (Coulomb) model. Horizontal lines are guides to the eyes, while the others are fits to A/N - with $A \sim 3$, a characteristic dependence on linear size of order parameter in the disordered phase.

Journal of Physics: Conference Series **686** (2016) 012005 doi:10.1088/1742-6596/686/1/012005

4. Conclusions and perspectives

We found special difficulty equilibrating large systems at low temperatures, due to the high degree of degeneracy and frustration of these models - notorious to present slow and arrested dynamics [41, 47, 48]. A first step towards a speed up in computational time would be to implement an extended ensemble Langevin dynamics, like parallel tempering for example. To further explore the properties of the nematic phases, it is important to measure the nematic correlation functions, as done in previous work [23]. In this sense, it is important to notice that the current estimation of the local director field makes use of a second order finite-difference scheme to calculate the gradient. This makes the director field susceptible to local fluctuations, and working with higher order stencils doesn't make significant improvement. As such, the correlations fall off substantially in the short scales, making the quantification of long-range properties difficult to analyse. The approach taken previously was to smooth out fluctuation by filters. In order not to damage the local orientational information of the configuration, there are specially designed filters that conserve the directional properties of the stripes [49]. Another interesting approach is to estimate the director field with the aid of the gradient-square tensor in Fourier space [50]. Finally, we emphasize the positive general features of the Langevin simulation technique to study such pattern forming systems where isotropy is guaranteed by careful designed conditions (see Figure 6).

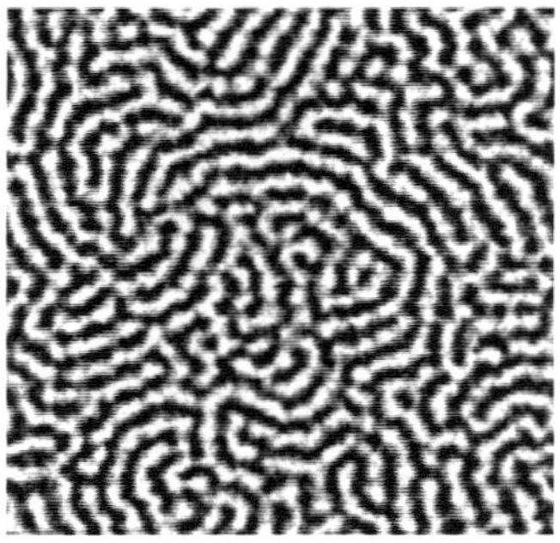

Figure 6. A typical $(42 \times 11)^2$ configuration of the isotropic phase for the dipolar model, close to the phase transition at $T = 0.60$.

We have presented results confirming the effect of the range of the competing isotropic interactions at the nature of the purely orientational low-temperature stripe phase, the nematic phase. In the dipolar interaction case orientational order parameter susceptibility seems to be divergent in the low temperature phase, characteristic of a critical phase as in the KT transition of XY model. Conversely, in the Coulomb interaction case we observed a saturation of this susceptibility, intrinsic to the broken symmetry phases of second order phase transitions. The similar behaviour for both cases close to the phase transition is in accordance with the reminiscent KT behaviour in equation (4). We expect to further explore the finite size properties of this distinctive behaviour in future extended simulations.

Acknowledgements
We gratefully acknowledge partial financial support from CNPq (Brazil) and the *Laboratório de Física Computacional* from IF-UFRGS for the utilization of the cluster Ada.

References

[1] Mendoza-Coto A, Stariolo D A and Nicolao L 2015 *Phys. Rev. Lett.* **114**(11) 116101

[2] Fischer K H and Hertz J A 1991 *Spin Glasses* (Cambridge: Cambridge University Press)
[3] Hubert A and Schafer R 1998 *Magnetic Domains* (Berlin: Springer-Verlag)
[4] Vega D A, Harrison C K, Angelescu D E, Trawick M L, Huse D A, Chaikin P M and Register R A 2005 *Phys. Rev. E* **71** 061803
[5] Kivelson S A, Fradkin E and Emery V J 1998 *Nature* **393** 550
[6] Vaterlaus A, Stamm C, Maier U, Pini M G, Politi P and Pescia D 2000 *Phys. Rev. Lett.* **84** 2247–2250
[7] Brazovskii S A 1975 *Sov. Phys. JETP* **41** 85–89
[8] Swift J and Hohenberg P C 1977 *Physical Review A* **15** 319
[9] Emery V and Kivelson S 1993 *Physica C: Superconductivity* **209** 597 – 621 ISSN 0921-4534
[10] Fradkin E and Kivelson S A 1999 *Phys. Rev. B* **59** 8065–8072
[11] Fradkin E, Kivelson S A, Lawler M J, Eisenstein J P and Mackenzie A P 2010 *Annual Review of Condensed Matter Physics* **1** 153–178
[12] Kivelson S A, Bindloss I P, Fradkin E, Oganesyan V, Tranquada J M, Kapitulnik A and Howald C 2003 *Rev. Mod. Phys.* **75**(4) 1201–1241
[13] Vojta M 2009 *Advances in Physics* **58** 699–820
[14] Fradkin E, Kivelson S A and Tranquada J M 2015 *Rev. Mod. Phys.* **87**(2) 457–482
[15] Strandburg K J 1988 *Rev. Mod. Phys* **60** 161–206
[16] Seul M and Andelman D 1995 *Science* **267** 476–483
[17] Wu Y Z, Won C, Scholl A, Doran A, Zhao H W, Jin X F and Qiu Z Q 2004 *Physical Review Letters* **93** 117205
[18] Abu-Libdeh N and Venus D 2009 *Phys. Rev. B* **80** 184412
[19] HAN J, WANG Q H and LEE D H 2001 *International Journal of Modern Physics B* **15** 1117–1126
[20] Borzi R A, Grigera S A, Farrell J, Perry R S, Lister S J S, Lee S L, Tennant D A, Maeno Y and Mackenzie A P 2007 *Science* **315** 214–217
[21] Parker C V, Aynajian P, da Silva Neto E H, Pushp A, Ono S, Wen J, Xu Z, Gu G and Yazdani A 2010 *Nature* **468** 677–680
[22] Barci D G and Stariolo D A 2007 *Physical Review Letters* **98** 200604
[23] Nicolao L and Stariolo D A 2007 *Phys. Rev. B* . **76** 054453
[24] Barci D G and Stariolo D A 2009 *Physical Review B* **79** 075437
[25] Toner J and Nelson D R 1981 *Phys. Rev. B* **23** 316–334
[26] Maier P G and Schwabl F 2004 *Phys. Rev. B* **70**(13) 134430
[27] Löw U, Emery V J, Fabricius K and Kivelson S A 1994 *Phys. Rev. Lett.* **72**(12) 1918–1921
[28] MacIsaac A B, Whitehead J P, Robinson M C and De'Bell K 1995 *Phys. Rev. B* **51** 16033
[29] Cannas S A, Michelon M F, Stariolo D A and Tamarit F A 2006 *Phys. Rev. B* **73** 184425
[30] Cannas S A, Michelon M F, Stariolo D A and Tamarit F A 2008 *Phys. Rev. E* **78** 051602
[31] Gleiser P M, Tamarit F A, Cannas S A and Montemurro M A 2003 *Phys. Rev. B* **68** 134401
[32] Abanov A, Kalatsky V, Pokrovsky V L and Saslow W M 1995 *Phys. Rev. B* **51** 1023–1038
[33] Cooper F, Freedman B and Preston D 1982 *Nuclear Physics B* **210** 210 – 228 ISSN 0550-3213
[34] Milchev A, Heermann D and Binder K 1986 *Journal of Statistical Physics* **44** 749–784 ISSN 0022-4715
[35] Davies C, Batrouni G, Katz G, Kronfeld A, Lepage P, Rossi P, Svetitsky B and Wilson K 1986 *Journal of Statistical Physics* **43** 1073–1075 ISSN 0022-4715
[36] Ukawa A 2015 *Journal of Statistical Physics* **160** 1081–1124 ISSN 0022-4715
[37] Langer J S, Bar-on M and Miller H D 1975 *Phys. Rev. A* **11**(4) 1417–1429
[38] Bray A J 1994 *Advances in Physics* **43** 357
[39] Roland C and Desai R 1990 *Phys. Rev. B* **42** 6658
[40] Hohenberg P C and Halperin B I 1977 *Rev. Mod. Phys.* **49** 435–479
[41] Daz-Mndez R, Mendoza-Coto A, Mulet R, Nicolao L and Stariolo D 2011 *The European Physical Journal B* **81**(3) 309–319 ISSN 1434-6028
[42] José J V, Kadanoff L P, Kirkpatrick S and Nelson D R 1977 *Phys. Rev. B* **16**(3) 1217–1241
[43] Taroni A, Bramwell S T and Holdsworth P C W 2008 *Journal of Physics: Condensed Matter* **20** 275233
[44] Jagla E A 2004 *Phys. Rev. E* **70** 046204
[45] Vollmayr-Lee B P and Rutenberg A D 2003 *Phys. Rev. E* **68**(6) 066703
[46] Frigo M and Johnson S 2005 *Proceedings of the IEEE* **93** 216–231 ISSN 0018-9219
[47] Ribeiro Teixeira A C, Stariolo D A and Barci D G 2013 *Phys. Rev. E* **87**(6) 062121
[48] Riesch C, Radons G and Magerle R 2014 *Phys. Rev. E* **90**(5) 052101
[49] Harrison C, Cheng Z, Sethuraman S, Huse D A, Chaikin P M, Vega D A, Sebastian J M, Register R A and Adamson D H 2002 *Phys. Rev. E* **66**(1) 011706
[50] van de Weijer J, van Vliet L, Verbeek P and van Ginkel R 2001 *Pattern Analysis and Machine Intelligence, IEEE Transactions on* **23** 1035–1042 ISSN 0162-8828

Journal of Physics: Conference Series **686** (2016) 012006 doi:10.1088/1742-6596/686/1/012006

Criticality and Chaos in Systems of Communities

Massimo Ostilli and Wagner Figueiredo

Departamento de Fisica, Universidade Federal de Santa Catarina, Florianopolis, 88040-900, SC, Brazil

E-mail: massimo.ostilli@gmail.com

Abstract. We consider a simple model of communities interacting via bilinear terms. After analyzing the thermal equilibrium case, which can be described by an Hamiltonian, we introduce the dynamics that, for Ising-like variables, reduces to a Glauber-like dynamics. We analyze and compare four different versions of the dynamics: flow (differential equations), map (discrete-time dynamics), local-time update flow, and local-time update map. The presence of only bilinear interactions prevent the flow cases to develop any dynamical instability, the system converging always to the thermal equilibrium. The situation is different for the map when unfriendly couplings are involved, where period-two oscillations arise. In the case of the map with local-time updates, oscillations of any period and chaos can arise as a consequence of the reciprocal "tension" accumulated among the communities during their sleeping time interval. The resulting chaos can be of two kinds: true chaos characterized by positive Lyapunov exponent and bifurcation cascades, or marginal chaos characterized by zero Lyapunov exponent and critical continuous regions.

1. Introduction

The fact that the speed of physical interactions is finite, as for light or sound, has well known crucial consequences in all physics, and it is the ultimate reason for the existence of waves. For example, if we artificially assume the speed of light as infinite in the Maxwell equations, the magnetic field would be zero and we would not have electromagnetic waves.

The situation is quite different when the equations contain non linear terms, which can result in erratic or even chaotic trajectories. Notice that, in the absence of mechanical kinetic terms, which is our framework, the presence of non linear terms is exclusively due to the presence of non bilinear interactions among the components of the system. For example, in the Kuramoto model [6]-[10], sinusoidal interactions lead to chaos [2]-[5] if the coupling constant K is larger than some threshold K_c.

It should be emphasized that the presence of non linear terms is not always sufficient for the emergence of chaos. However, it has been observed that the introduction of time-delays among the components of a system augments the chance that the system develops chaos [9]-[13]. The idea at the base of such a mechanism is that time-delays among the system components makes it difficult for the system to set in a steady state. In particular, for linear flows, a fixed delay among the system components corresponds to a finite speed of interactions, which in turn leads to oscillations or waves. For non linear flows a delay can have more dramatic consequences, with chaos scenarios more probable.

In [17], we have introduced a simple discrete-time two components dynamical model, characterized by local-time updates. Local time updates means that each component updates its status only at certain times t. In the specific model analyzed in [17], a component updates (sleeps) its status only at odd times $t = 1, 3, \ldots$, (even times $t = 0, 2, \ldots$), while the other component updates (sleeps) its status only at even times $t = 0, 2, \ldots$ (odd times $t = 1, 3, \ldots$). Remarkably, as we have shown in [17], for certain coupling values, this simple system exhibits chaos even if it is characterized only by bilinear interactions. In fact, when the couplings are in the steady state regime, the system can be described via a Hamiltonian which is quadratic in the system components. We are not aware of a similar case in the literature. Our understanding here is that there are three factors generating chaos: *i)* the presence of unfriendly couplings, *ii)* the discrete-time nature of the dynamics, and *iii)* the existence of at least two interacting communities which rearrange their configurations at alternate times, *i.e.* via local-time updates. Notice that *(iii)* is not just a delay; it also implies that each component has a sleeping period. It is during such a sleeping period that the component accumulates a tension with respect to the other component when the reciprocal coupling is unfriendly. We are in fact dealing with a sort of dynamical frustration.

In this paper we review some aspects of the model which were only partially treated in [17]. In particular, we compare the analogous flow model with the present map and show that the above conditions *(i)-(iii)* are in fact necessary, and calculate the maximal Lyapunov exponent.

The reader not interested in the derivation of the dynamic equations may skip the next Section.

2. The model in the flow, map and local-time versions

Here we consider a two components system. Generalization to an arbitrary number of components will be reported elsewhere. We consider two communities [14] of agents $\mathcal{N}^{(1)}$ and $\mathcal{N}^{(2)}$, with cardinalities $N^{(1)}$ and $N^{(2)}$. The agent with index i can be in two possible status, $\sigma_i = \pm 1$, with $i \in \mathcal{N}^{(1)} \cup \mathcal{N}^{(2)}$. According to the sign of the couplings, friendly or unfriendly, each agent i tends to follow or anti-follow its neighbors, by minimizing or maximizing the term $\sigma_i \sum' \sigma_j$, where $\sum'$ runs over the set of neighbors of i. It is necessary to distinguish between intra- and inter-couplings, therefore we introduce the 2×2 matrix $J^{(l,m)}$: $J^{(1,1)}$ and $J^{(2,2)}$ are the intra-couplings, and $J^{(1,2)} = J^{(2,1)}$ the inter-coupling. In the most general formulation we should introduce also the 2×2 matrix $\Gamma^{(l,m)}$ defined as the set of coupled spins (i, j) within the same community (intra), or between the two communities (inter). Finally, we introduce a global factor β that rescales all the couplings in $\beta J^{(l,m)}$.

Let us indicate the time by t, which is supposed to range in the set of natural numbers $\mathbb{N}$ for the map, while it can take any non negative real number for the flow. Inspired by the Glauber dynamics [18], for flow and map we introduce the following transition rate probabilities for the spin with state σ_i to jump to the state σ_i'

$$w^{(\mathrm{FLOW/MAP})}(\sigma_i \to \sigma_i'; t) = \begin{cases} \frac{1+\sigma_i' \tanh\left(\beta J^{(1,1)} \sum_{j:(i,j)\in\Gamma^{(1,1)}} \sigma_j + \beta J^{(1,2)} \sum_{j:(i,j)\in\Gamma^{(1,2)}} \sigma_j\right)}{2}, & \text{for } i \in \mathcal{N}^{(1)}, \\ \frac{1+\sigma_i' \tanh\left(\beta J^{(2,2)} \sum_{j:(i,j)\in\Gamma^{(2,2)}} \sigma_j + \beta J^{(2,1)} \sum_{j:(i,j)\in\Gamma^{(2,1)}} \sigma_j\right)}{2}, & \text{for } i \in \mathcal{N}^{(2)}. \end{cases} \quad (1)$$

Let $(\mathbb{E}, \mathbb{O})$ be any periodic partition of $\mathbb{N}$, *i.e.*, $\mathbb{E} \cup \mathbb{O} = \mathbb{N}$, $\mathbb{E} \cap \mathbb{O} = \emptyset$, and $\mathbb{E}$ and $\mathbb{O}$ contain both infinite elements of $\mathbb{N}$. For instance, $\mathbb{E}$ and $\mathbb{O}$ can be the set of even and odd numbers, respectively, and we will assume this choice in all the next examples. We now introduce the following local-time transition rate probabilities for the spin with state σ_i to jump to the state σ_i'

$$w^{(\mathrm{LOCAL-TIME\ FLOW})}(\sigma_i \to \sigma_i'; t) = \begin{cases} \delta_{\sigma_i,\sigma_i'} \quad i \in \mathcal{N}^{(1)}, \quad \text{if } \mathrm{mod}[t,2] \neq 0, \\ \frac{1+\sigma_i' \tanh\left(\beta J^{(1,1)} \sum_{j:(i,j)\in\Gamma^{(1,1)}} \sigma_j + \beta J^{(1,2)} \sum_{j:(i,j)\in\Gamma^{(1,2)}} \sigma_j\right)}{2}, \\ \text{for } i \in \mathcal{N}^{(1)}, \quad \text{if } \mathrm{mod}[t,2] = 0, \\ \delta_{\sigma_i,\sigma_i'} \quad i \in \mathcal{N}^{(2)}, \quad \text{if } \mathrm{mod}[t,2] \neq 0, \\ \frac{1+\sigma_i' \tanh\left(\beta J^{(2,2)} \sum_{j:(i,j)\in\Gamma^{(2,2)}} \sigma_j + \beta J^{(2,1)} \sum_{j:(i,j)\in\Gamma^{(2,1)}} \sigma_j\right)}{2}, \\ \text{for } i \in \mathcal{N}^{(2)}, \quad \text{if } \mathrm{mod}[t,2] = 0, \end{cases} \tag{2}$$

$$w^{(\mathrm{LOCAL-TIME\ MAP})}(\sigma_i \to \sigma_i'; t) = \begin{cases} \delta_{\sigma_i,\sigma_i'} \quad i \in \mathcal{N}^{(1)}, \quad \text{for } t \in \mathbb{O}, \\ \frac{1+\sigma_i' \tanh\left(\beta J^{(1,1)} \sum_{j:(i,j)\in\Gamma^{(1,1)}} \sigma_j + \beta J^{(1,2)} \sum_{j:(i,j)\in\Gamma^{(1,2)}} \sigma_j\right)}{2}, \\ \text{for } i \in \mathcal{N}^{(1)}, \quad t \in \mathbb{E}, \\ \delta_{\sigma_i,\sigma_i'} \quad i \in \mathcal{N}^{(2)}, \quad \text{for } t \in \mathbb{E}, \\ \frac{1+\sigma_i' \tanh\left(\beta J^{(2,2)} \sum_{j:(i,j)\in\Gamma^{(2,2)}} \sigma_j + \beta J^{(2,1)} \sum_{j:(i,j)\in\Gamma^{(2,1)}} \sigma_j\right)}{2}, \\ \text{for } i \in \mathcal{N}^{(2)}, \quad t \in \mathbb{O}. \end{cases} \tag{3}$$

From the viewpoint of modeling, Eqs. (1), (2), or (3), are justified as they make each spin to follow the majority of its intra- and inter-neighbors and, thanks to the presence of the functions $\tanh(\cdot)$, the rates are non negative and normalized at any time $\sum_{\sigma'} w(\sigma \to \sigma'; t) = 1/(\text{Time Unit})$. From a deeper viewpoint, Eqs. (1), (2), or (3), are based on the fact that, as we will see soon, in the case of positive couplings they lead to Boltzmann equilibrium governed by only quadratic interactions. More precisely, in the case of positive couplings these forms guarantee that at equilibrium the system satisfies the principle of detailed balance and the principle of maximal entropy for any quadratic interactions.

We formalize the discrete-time probabilistic dynamics induced by Eqs. (1), (2), or (3), as follows. Let $N = N^{(1)} + N^{(2)}$. Let us introduce the spin vector $\boldsymbol{\sigma} = (\sigma_1, \ldots, \sigma_N)$, and the associated probability vector $p(\boldsymbol{\sigma}; t)$, *i.e.*, the probability that the system is in the configuration $\boldsymbol{\sigma}$ at time $t \in \mathbb{N}$. The master equation for flow and map (local-time or not) reads

$$\frac{\partial p^{(\mathrm{FLOW})}(\boldsymbol{\sigma}; t)}{\partial(\alpha t)} = -\sum_{\boldsymbol{\sigma}'} p^{(\mathrm{FLOW})}(\boldsymbol{\sigma}; t) W(\boldsymbol{\sigma} \to \boldsymbol{\sigma}') + \sum_{\boldsymbol{\sigma}'} p^{(\mathrm{FLOW})}(\boldsymbol{\sigma}'; t) W(\boldsymbol{\sigma}' \to \boldsymbol{\sigma}), \tag{4}$$

$$\frac{p^{(\mathrm{MAP})}(\boldsymbol{\sigma}; t+1) - p^{(\mathrm{MAP})}(\boldsymbol{\sigma}; t)}{\alpha} = -\sum_{\boldsymbol{\sigma}'} p^{(\mathrm{MAP})}(\boldsymbol{\sigma}; t) W(\boldsymbol{\sigma} \to \boldsymbol{\sigma}') + \sum_{\boldsymbol{\sigma}'} p^{(\mathrm{MAP})}(\boldsymbol{\sigma}'; t) W(\boldsymbol{\sigma}' \to \boldsymbol{\sigma}), \tag{5}$$

where we have introduced the global transition rates

$$W(\boldsymbol{\sigma} \to \boldsymbol{\sigma}') = \prod_{i \in \mathcal{N}^{(1)} \cup \mathcal{N}^{(2)}} w(\sigma_i \to \sigma_i'), \tag{6}$$

Journal of Physics: Conference Series **686** (2016) 012006 doi:10.1088/1742-6596/686/1/012006

where $w(\sigma_i \to \sigma_i')$ can be of the type Eqs. (1), (2), or (3), and where $\alpha/2 > 0$ may be interpreted as the rate at which, due to the interaction with an environment, a free spin ($J^{(l,m)} = 0$) makes transitions from either state to the other. As we have proved in [15], it is necessary to impose the bound $\alpha \le 1$ for $p(\boldsymbol{\sigma}; t)$ to be a probability at any time t. By using Eqs. (2)-(6) it is easy to check that the stationary solutions $p(\boldsymbol{\sigma})$ of Eq. (5) are given by the Boltzmann distribution $p(\boldsymbol{\sigma}) \propto \exp[-\beta H(\boldsymbol{\sigma})]$, where

$$H = -J^{(1,1)} \sum_{(i,j)\in\Gamma^{(1,1)}} \sigma_i\sigma_j - J^{(2,2)} \sum_{(i,j)\in\Gamma^{(2,2)}} \sigma_i\sigma_j - J^{(1,2)} \sum_{(i,j)\in\Gamma^{(1,2)}} \sigma_i\sigma_j. \qquad (7)$$

Of course, the existence of a stationary solution $p(\boldsymbol{\sigma})$ does not represent a sufficient condition for equilibrium. In fact, asymptotically the system can reach non-point-like attractors, and even aperiodic or chaotic regimes. However, as we have discussed in the Introduction, the fact that for certain values of couplings the system can be described via the Hamiltonian (7), means that we are dealing with only bilinear interactions. Here, the non linearities that lead to chaos are caused by the local time updates.

From now on, we will omit the suffixes (FLOW), (MAP) and (LOCAL-TIME). It will be clear from the context what we are referring to. Eqs. (2)-(6) define the microscopic dynamics from which one can derive the macroscopic dynamics, *i.e.*, the dynamics for the order parameters

$$x^{(1)}(t) = \sum_{\boldsymbol{\sigma}} p(\boldsymbol{\sigma}; t) \frac{1}{N} \sum_{i\in\mathcal{N}^{(1)}} \sigma_i, \qquad (8)$$

$$x^{(2)}(t) = \sum_{\boldsymbol{\sigma}} p(\boldsymbol{\sigma}; t) \frac{1}{N} \sum_{i\in\mathcal{N}^{(2)}} \sigma_i. \qquad (9)$$

3. Mean Field Limit

The mean-field limit is defined by the settings $|\Gamma^{(1,1)}| = \binom{N^{(1)}}{2}$, $|\Gamma^{(2,2)}| = \binom{N^{(2)}}{2}$, $|\Gamma^{(1,2)}| = N^{(1)}N^{(2)}$, and the replacements $J^{(1,1)} \to J^{(1,1)}/N^{(1)}$, $J^{(2,2)} \to J^{(2,2)}/N^{(2)}$, $J^{(1,2)} \to J^{(1,2)}(N^{(1)} + N^{(2)})/(2N^{(1)}N^{(2)})$. We parametrize the size of the two communities as

$$N^{(1)} = N\rho^{(1)}, \quad N^{(2)} = N\rho^{(2)}, \quad \rho^{(1)} + \rho^{(2)} = 1. \qquad (10)$$

Let us introduce the matrix $\tilde{\boldsymbol{J}}$

$$\tilde{\boldsymbol{J}} = \begin{pmatrix} J^{(1,1)} & \frac{J^{(1,2)}}{2\rho^{(1)}} \\ \frac{J^{(2,1)}}{2\rho^{(2)}} & J^{(2,2)} \end{pmatrix}. \qquad (11)$$

As shown in [17], for $N \to \infty$ we obtain the following deterministic evolution Eqs. for the order parameters (8)-(9)

3.1. FLOW

$$\frac{\partial x^{(1)}(t)}{\partial(\alpha t)} = \tanh\left(\beta \tilde{J}^{(1,1)} x^{(1)}(t) + \beta \tilde{J}^{(1,2)} x^{(2)}(t)\right) - x^{(1)}(t), \qquad (12)$$

$$\frac{\partial x^{(2)}(t)}{\partial(\alpha t)} = \tanh\left(\beta \tilde{J}^{(2,2)} x^{(2)}(t) + \beta \tilde{J}^{(2,1)} x^{(1)}(t)\right) - x^{(2)}(t), \qquad (13)$$

3.2. MAP

$$\frac{x^{(1)}(t+1)-x^{(1)}(t)}{\alpha} = \tanh\left(\beta\tilde{J}^{(1,1)}x^{(1)}(t)+\beta\tilde{J}^{(1,2)}x^{(2)}(t)\right)-x^{(1)}(t), \tag{14}$$

$$\frac{x^{(2)}(t+1)-x^{(2)}(t)}{\alpha} = \tanh\left(\beta\tilde{J}^{(2,2)}x^{(2)}(t)+\beta\tilde{J}^{(2,1)}x^{(1)}(t)\right)-x^{(2)}(t), \tag{15}$$

3.3. LOCAL-TIME FLOW

$$\frac{\partial x^{(1)}(t)}{\partial(\alpha t)} = \begin{cases} 0, \quad \text{if } \mod[t,2]\neq 0, \\ \tanh\left(\beta\tilde{J}^{(1,1)}x^{(1)}(t)+\beta\tilde{J}^{(1,2)}x^{(2)}(t)\right)-x^{(1)}(t), \ \text{if } \mod[t,2]=0, \end{cases} \tag{16}$$

$$\frac{\partial x^{(2)}(t)}{\partial(\alpha t)} = \begin{cases} 0, \quad \text{if } \mod[t,2]\neq 0, \\ \tanh\left(\beta\tilde{J}^{(2,2)}x^{(2)}(t)+\beta\tilde{J}^{(2,1)}x^{(1)}(t)\right)-x^{(2)}(t), \ \text{if } \mod[t,2]=0. \end{cases} \tag{17}$$

3.4. LOCAL-TIME MAP

$$\frac{x^{(1)}(t+1)-x^{(1)}(t)}{\alpha} = \begin{cases} 0, \quad t\in\mathbb{O}, \\ \tanh\left(\beta\tilde{J}^{(1,1)}x^{(1)}(t)+\beta\tilde{J}^{(1,2)}x^{(2)}(t)\right)-x^{(1)}(t), \ t\in\mathbb{E}, \end{cases} \tag{18}$$

$$\frac{x^{(2)}(t+1)-x^{(2)}(t)}{\alpha} = \begin{cases} 0, \quad t\in\mathbb{E}, \\ \tanh\left(\beta\tilde{J}^{(2,2)}x^{(2)}(t)+\beta\tilde{J}^{(2,1)}x^{(1)}(t)\right)-x^{(2)}(t), \ t\in\mathbb{O}. \end{cases} \tag{19}$$

3.5. Stationary solutions

When the couplings are positive, there is little difference among the four dynamics, and they all asymptotically reach equilibrium according to the Boltzmann distribution $p(\boldsymbol{\sigma}) \propto \exp[-\beta H_{\mathrm{mf}}(\boldsymbol{\sigma})]$, where $H_{\mathrm{mf}}(\boldsymbol{\sigma})]$ is the mean-field analogous of Eq. (7), and $x^{(1)}(t)$ and $x^{(2)}(t)$ tend, for $t\to\infty$, to the stationary solutions of Eqs. (14)-(17), *i.e.*,

$$\begin{cases} x^{(1)} = \tanh\left(\beta\tilde{J}^{(1,1)}x^{(1)}+\beta\tilde{J}^{(1,2)}x^{(2)}\right), \\ x^{(2)} = \tanh\left(\beta\tilde{J}^{(2,1)}x^{(1)}+\beta\tilde{J}^{(2,2)}x^{(2)}\right). \end{cases} \tag{20}$$

Eqs. (20) represent a particular case of the general result derived in [16] valid for n interacting communities at equilibrium. In particular, one can check that Eqs. (20) give rise to second order phase transitions whose critical surface is determined by the condition

$$\det\left(\mathbf{1}-\beta\tilde{\boldsymbol{J}}\right)=0. \tag{21}$$

In general, $(x^{(1)},x^{(2)})=(0,0)$ is stable when the eigenvalues of $\beta\tilde{\boldsymbol{J}}$ are inside the interval $(-1,1)$ (disordered phase, or no consensus), otherwise the system reaches a spontaneous ordered phase $(x^{(1)},x^{(2)})\neq(0,0)$ (frozen phase, or consensus). However, Eq. (21) is exact only for the flow and the map. In general, due to the sleeping time, the critical value of β can differ from the one given by Eq. (21). This was erroneously not stressed in [17]. The correct way to to derive the exact critical value of β must use the full expression of the Jacobian, which is the subject of Section 5.

4. Comparing flow, map and local-time versions

From now on we shall always assume $\alpha = 1$ and $N^{(1)} = N^{(2)}$.

In this Section we compare the four dynamics numerically.

In Figs (1a)-(2b) we compare the normal flow and map. In Figs. (1a) and (1b) we consider a situation with positive couplings at small and large values of β; in Figs. (2a) and (2b) we consider a situation with positive and negative couplings at small and large values of β, respectively. It is evident that, if the couplings are positive, the flow and the map remain roughly close at any time, and tend to coincide for $t \to \infty$ to the same stationary solution provided by Eq. (20). When instead some of the couplings are negative, at small β the map tends to make damping oscillations around the flow (which for the present case is null), whereas, at large β, the map tends to make regular oscillations around the flow solution (which for the present case is null).

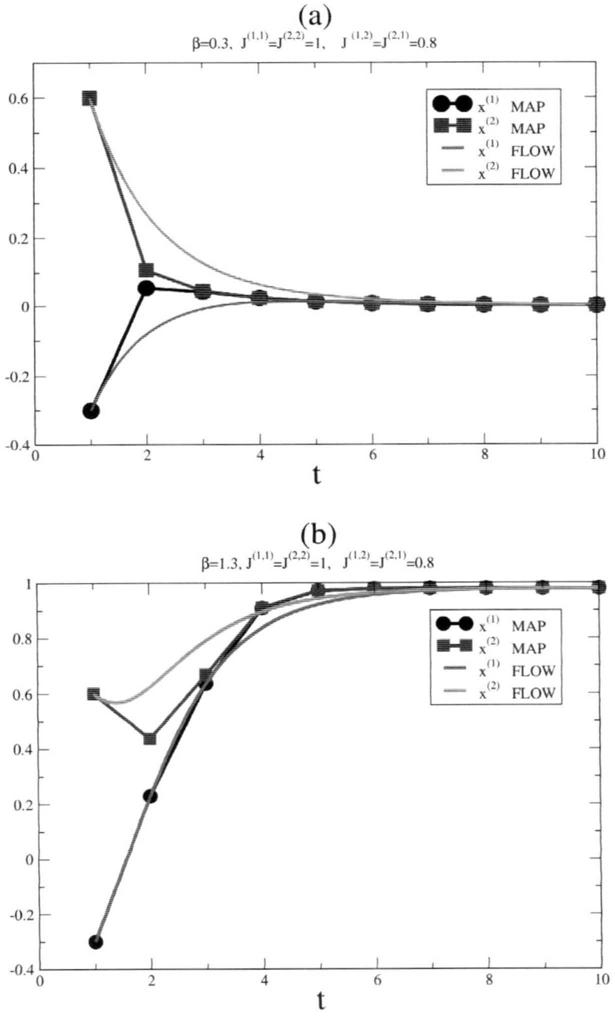

Figure 1: (Color online) Plot of $x^{(1)}(t)$ and $x^{(2)}(t)$, solution of the flow, Eqs. (12)-(13), and the map, Eqs. (14)-(15) (here with $\alpha = 1$), for positive couplings; below (Panel a) and above (Panel b) the threshold β_c. Initial conditions are the same in all cases. The lines of the map are guides for the eyes.

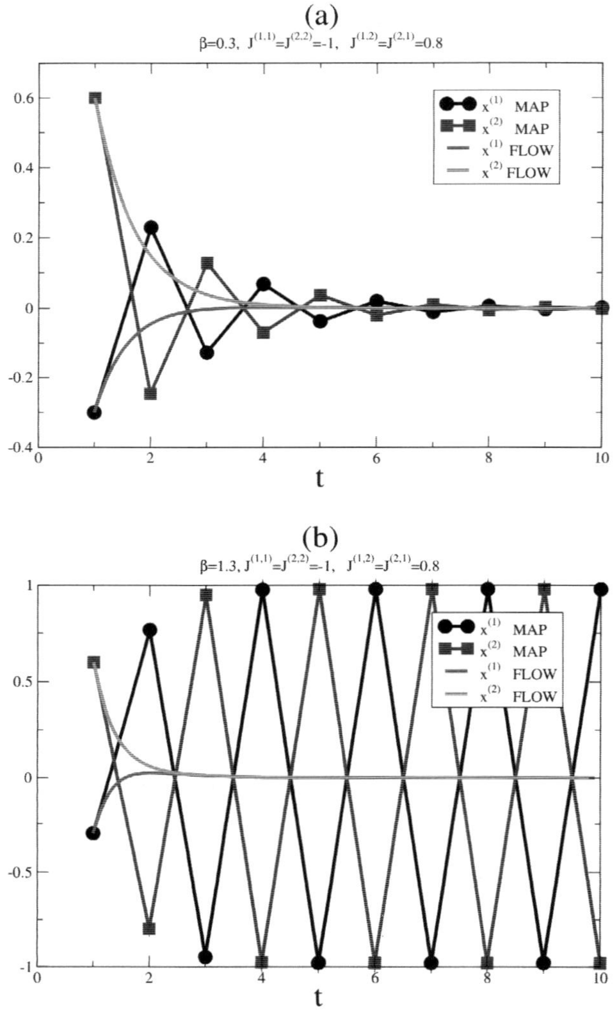

Figure 2: (Color online) Plot of $x^{(1)}(t)$ and $x^{(2)}(t)$, solution of the flow, Eqs. (12)-(13), and the map, Eqs. (14)-(15) (here with $\alpha = 1$), for a case with positive and negative couplings; below (Panel a) and above (Panel b) the threshold β_c. Initial conditions are the same in all cases. The lines of the map are guides for the eyes.

In Figs. (3a)-(3c) we compare the flow and map with local-time updates for a case with positive and negative couplings and three values of β, respectively. Fig. (3a) provides the behavior for $\beta = 1.5$ where the map follows damping oscillations and both the map and the flow tend to the same stationary value provided by Eq. (20). Fig. (3b) provides the behavior for a relatively larger value of β, $\beta = 2.5$, where the map follows regular oscillations of period 2, while the flow tends to a stationary value. Fig. (3c) provides the behavior for $\beta = 4.5$, where the map follows regular oscillations of period 4, while the flow tends to a stationary value. Increasing further β leads to larger and larger periods (not shown).

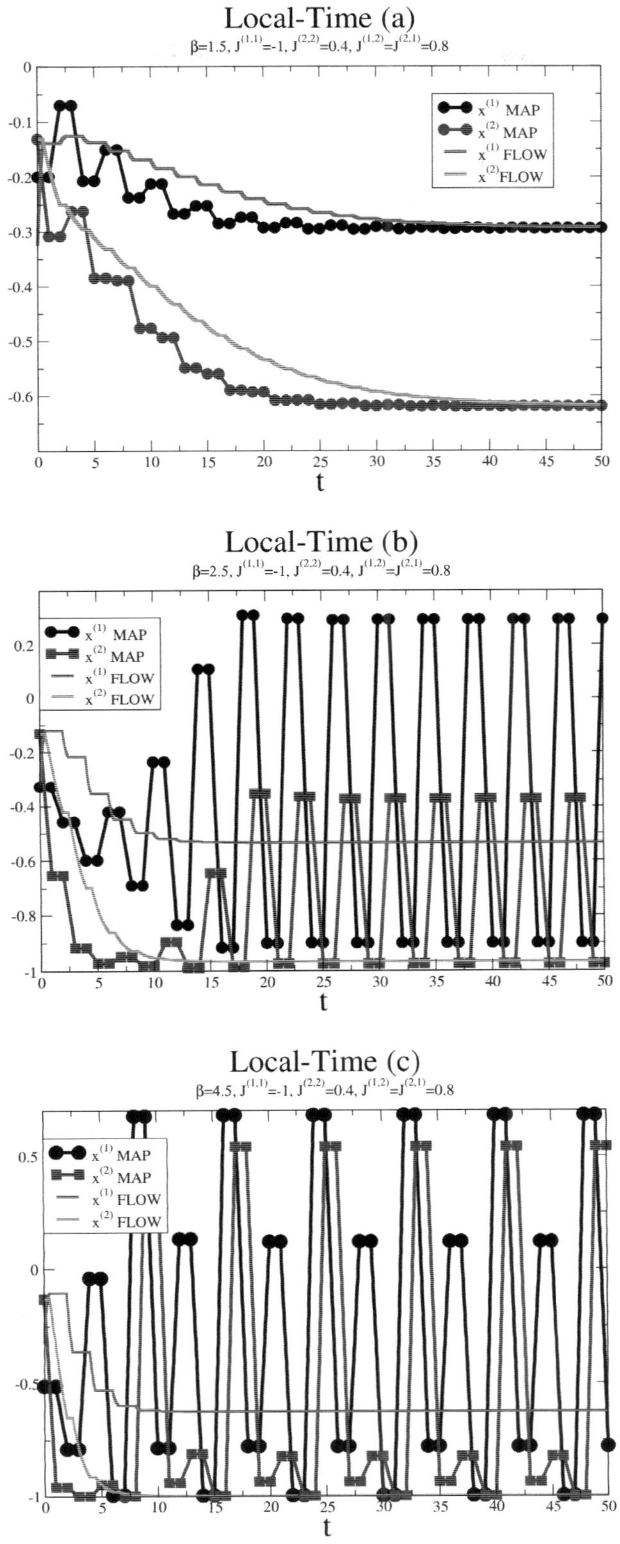

Figure 3: (Color online) Plot of $x^{(1)}(t)$ and $x^{(2)}(t)$, solution of the local-time flow, Eqs. (16)-(17), and the local-time map, Eqs. (18)-(19) (here with $\alpha = 1$), $\beta = 1.5$ (a), $\beta = 2.5$ (b), and $\beta = 4.5$ (c). Initial conditions are the same in all cases. The lines of the map are guides for the eyes.

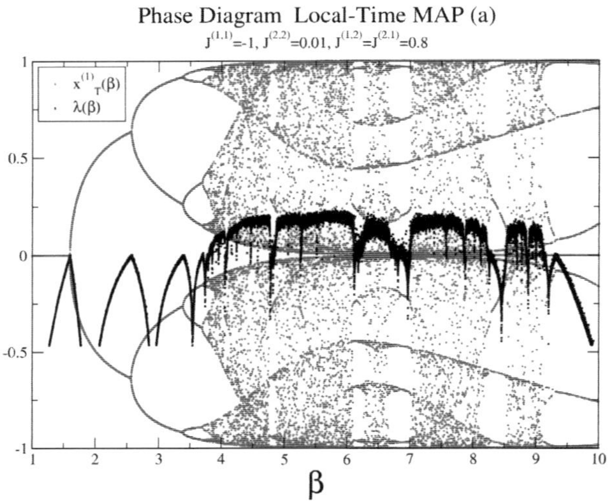

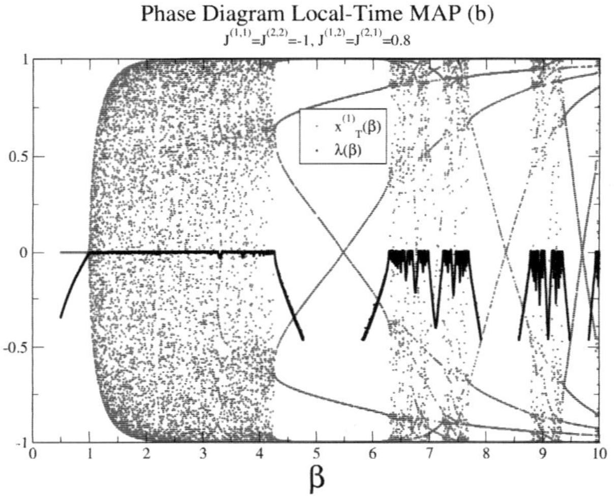

Figure 4: (Color online) Plot of $x_T^{(1)}(\beta)$ (red dots), solution of the local-time map, Eqs. (18)-(19) (here with $\alpha = 1$) as a function of β, with T high enough to remove temporary transients. In this case $T = 10^4$. We plot also the maximum Lyapunov exponent $\lambda(\beta)$ (black highly irregular line) calculated from Eqs. (25)-(27). Each value of β is in correspondence with a random initial condition for a total of 10^5 samples. Panel (a) is a case with bifurcation cascade, Panel (b) is a case with marginal chaos.

5. Chaos

The case analyzed in Figs. (3) shows that, for large values of β, the local-time map can lead to erratic and, possibly, chaotic trajectories. In general, for each dynamics, the major quantity of interest is the Jacobian $\mathcal{J}_t^{(l,m)}$, where $l, m \in \{1, 2\}$:

5.1. *FLOW*

$$\mathcal{J}_t^{(l,m)} = \left[1 - \left(x_t^{(l)}\right)^2\right] \beta J^{(l,m)}, \tag{22}$$

5.2. MAP

$$\mathcal{J}_{t+1}^{(l,m)} = \left[1 - \left(x_t^{(l)}\right)^2\right] \beta J^{(l,m)}, \tag{23}$$

5.3. LOCAL-TIME FLOW

$$\mathcal{J}_t^{(l,m)} = \begin{cases} \delta_{l,m}, & \mathrm{mod}[t+l-1,2] \neq 0, \\ \left[1 - \left(x_t^{(l)}\right)^2\right] \beta J^{(l,m)}, & \mathrm{mod}[t+l-1,2] = 0, \end{cases} \tag{24}$$

5.4. LOCAL-TIME MAP

$$\mathcal{J}_{t+1}^{(l,m)} = \begin{cases} \delta_{l,m}, & \mathrm{mod}[t+l-1,2] \neq 0, \\ \left[1 - \left(x_t^{(l)}\right)^2\right] \beta J^{(l,m)}, & \mathrm{mod}[t+l-1,2] = 0. \end{cases} \tag{25}$$

The knowledge of $\mathcal{J}_t^{(l,m)}$ allows to calculate the maximum Lyapunov exponent λ. It turns out that in the map case (local-time or not), due to the fact that - via recurrence - we have the explicit solution of the Eqs., the evaluation of λ is easier if compared to the flow case. Since the flow does not show any signs of chaos or erratic trajectories, we can limit the evaluation of the Lyapunov exponent to the map, where we can use

$$\lambda = \lim_{t\to\infty} \frac{1}{2t} \log(\Lambda_t), \tag{26}$$

where Λ_t is the largest eigenvalue of the following matrix

$$\boldsymbol{Q}_t = \boldsymbol{L}_t \boldsymbol{L}_t^{\dagger}, \quad \text{where} \quad \boldsymbol{L}_t = \prod_{t'=0}^{t} \boldsymbol{\mathcal{J}}_{t'}. \tag{27}$$

Given a set of couplings, we let the system to evolve toward high enough values of $t = T$ in order to remove temporary transients and repeat the numerical experiment for several values of β, each β being associated to a random initial condition. Hence, we plot $(x^{(1)}(T), x^{(2)}(T))$ as functions of β, and we indicate these functions as $(x_T^{(1)}(\beta), x_T^{(2)}(\beta))$. In our examples and for the range of β considered, $T \geq 10^3$ turns out to be enough high. In fact, the variable β plays the role of a time-scaling: the higher β, the higher the necessary transient T. In general, the functions $(x_T^{(1)}(\beta), x_T^{(2)}(\beta))$ look multi-valued functions due to the existence of bifurcation points. There exist two kinds of bifurcations: doubling period and phase transition. However, it is clear that in either case, doubling period or phase transition, the presence of bifurcations increases the chance to develop chaos. Therefore, for our aims here, it is more interesting, as well as highly more efficient, to show the plots that include all the bifurcation points. Figs. (4a)-(4b) show two different scenarios. The case of Panel (a) is a "classical" bifurcation cascade [1], where, after a threshold β_{Chaos}, bifurcation takes place increasingly, up to windows of stability. In the case of Panel (b) we observe a different situation which we call "marginal chaos". In this case, after a threshold $\beta_{Marginal}$, the dynamics is characterized by totally erratic trajectories with no period, up to windows of stability. Strictly speaking, even out of the windows of stability, the system is not chaotic, since the maximum Lyapunov exponent is $\lambda = 0$. However, in such regions, the system turns out to be critical over continuous intervals; a situation which is not typical in models characterized by discrete symmetries like the Ising model.

6. Conclusions

We have analyzed and compared four kinds of dynamics of a simple two component Ising-like system: flow, map, local-time flow, and local-time map. Remarkably, even if only bilinear interactions are present, the local-time map gives rise to erratic trajectories and, depending on the set of couplings, two chaotic scenarios take place: bifurcation cascades or "marginal chaos", *i.e.*, criticality extended over continuous intervals of β, the time-scale parameter. The analogous local-time flow does not present any of such behaviors, not even oscillations, confirming all the scenarios we have already discussed in [17]. We stress that this local-time update is not just a time-delay: here we have also a sleeping time during which, in the case of some unfriendly couplings, there is an accumulation of frustration among the system components, giving rise to instability and chaos, even for a system characterized by only bilinear interactions. As we have discussed in [17], we believe that maps with local-time updates are a quite common feature in complex systems. Urgent issues will be to understand how this scenario generalizes to large systems with many, say q components, how the probability to have chaos changes with q, and if the mean-field (or more precisely fully connected) picture remains robust.

6.1. Acknowledgments

Work supported by CNPq Grant PDS 150934/2013-0. W. F. also acknowledges the Brazilian agencies CNPq and CAPES.

References

[1] May M R 1976 *Simple mathematical models with very complicated dynamics*, Nature **261**, 459-467.
[2] Strogatz S H 1994 "Nonlinear Dynamics and Chaos, with Applications to Physics, Biology, Chemistry, and Engineering", Reading, MA: Addison-Wesley.
[3] Lorenz E N 1996 "The Essence of Chaos", Seattle, WA: University of Washington Press.
[4] Baker G L, and Gollub J B 1996 "Chaotic Dynamics: An Introduction", Cambridge.
[5] Krawiecki A, Hołyst J A , and Helbing D 2002 Phys. Rev. Lett. **89**, 158701.
[6] Kuramoto Y 1984 "Chemical Oscillations, Waves, and Turbulence", Springer-Verlag, New York, p. 164.
[7] Strogatz S H 2000 Physica D **143**, 1.
[8] Pikovsky A, Rosenblum M, and Kurths J 2001 *Synchronization, A Universal Concept in Nonlinear Sciences*, (Cambridge, UK: Cambridge University Press); D.Topaj, and Pikovsky A 2002 Physica D **170**, 118.
[9] Mackey D, Glass L 1977 Science **197**, 28.
[10] Kuramoto Y, Nakao H 1997 Phys. Rev. Lett. **21**, 4029.
[11] Mørk J, Tromborg B, and Mark J 1992 IEEE J. of Quantum Electronics, **28**, 93, No. 1.
[12] Stephen Yeung M K and Strogatz Steven H 1999 Phys. Rev. Lett, **28**, 648.
[13] Sprott J C 2007 Phys. Lett. A, **366**, 397.
[14] Fortunato S 2010 *Community detection in graphs*, Phys. Rep. **486**, 75.
[15] Ostilli M, Mukhamedov F 2013 Europhys. Lett. **101**, 60008.
[16] Ostilli M, Mendes J F 2009 Phys. Rev. E **80**, 011142.
[17] Ostilli M, Figueiredo W 2015 EPL **108**, 20009.
[18] Glauber R J 1963 J. Math. Phys. **4**, 294.

Journal of Physics: Conference Series **686** (2016) 012007 doi:10.1088/1742-6596/686/1/012007

Winding angle distributions for two-dimensional collapsing polymers

Arturo Narros

School of Mathematical Sciences, Queen Mary University of London, Mile End Road E1 4NS London, UK

Aleksander L Owczarek

School of Mathematics and Statistics, The University of Melbourne, Parkville, Vic 3010, Australia

Thomas Prellberg

School of Mathematical Sciences, Queen Mary University of London, Mile End Road E1 4NS London, UK

E-mail:
a.n.gonzalez@qmul.ac.uk,owczarek@unimelb.edu.au,thomas.prellberg@qmul.ac.uk

Abstract. We provide numerical support for a long-standing prediction of universal scaling of winding angle distributions. Simulations of interacting self-avoiding walks show that the winding angle distribution for N-step walks is compatible with the theoretical prediction of a Gaussian with a variance growing asymptotically as $C \log N$, with $C = 2$ in the swollen phase (previously verified), and $C = 24/7$ at the θ-point. At low temperatures weaker evidence demonstrates compatibility with the same scaling and a value of $C = 4$ in the collapsed phase, also as theoretically predicted.

1. Introduction

Polymers in a dilute solution can be either swollen or collapsed, or at an intermediate critical point, the so-called θ-point, depending on the quality of the solvent [1]. The canonical lattice model for polymer collapse is the interacting self-avoiding walk on a regular lattice such as the square or simple cubic lattice, in two or three spatial dimensions, respectively.

Some time ago, the winding angle distribution for polymers received theoretical attention. In two dimensions, the winding angle distribution of Brownian paths of length N satisfies Spitzer's law [2]

$$P(x = 2\theta/\log N) \sim \frac{1}{1+x^2} \,. \tag{1}$$

This distribution has divergent moments due to the infinitesimal winding centre. Changing to macroscopic winding centres, or, alternatively, to random walks on a lattice, changes the structure of the winding centre and leads to an exponential decay in the winding angle distribution. More precisely, the distribution gets altered to

$$P(x = 2\theta/\log N) \sim \frac{1}{\cosh^2(1+\pi x/2)} \tag{2}$$

for adsorbing winding centres and to

$$P(x = 2\theta/\log N) \sim \frac{1}{\cosh(1+\pi x/2)} \tag{3}$$

for reflecting winding centres [3, 4, 5].

The introduction of an excluded volume leads to a swelling of the polymer. Heuristically, swelling implies that the winding of segments $N/2$, $N/4$, $N/8$, ... become approximately independent. Hence an N-step chain has $O(\log N)$ segments and a law-of-large-numbers argument implies that the winding angle distribution becomes Gaussian with variance proportional to $\log N$. More precisely, one finds that

$$P(x = \theta/\sqrt{\log N}) \sim \exp(-x^2/(2C)). \tag{4}$$

Note that the introduction of the excluded volume has changed the scaling variable from $x = \theta/\log N$ to $x = \theta/\sqrt{\log N}$.

Using a mapping to a Coulomb gas [6], it has been argued that the constant C is universal in two dimensions, in that it assumes distinct values depending only on the phase the polymer is in. In particular, it has been predicted that

$$C = \begin{cases} 2 & \text{swollen phase,} \\ 24/7 & \text{critical state,} \\ 4 & \text{dense phase.} \end{cases} \tag{5}$$

This calculation has been extended to watermelon configurations in [7], where it was shown that the number L of legs in the watermelon configuration modifies the constant C by a factor $1/L^2$. As a consequence, it was argued that for θ-point polymers on the Manhattan lattice one finds $C = 6/7$, as the modification of the lattice leads to a choice of $L = 2$ instead of $L = 1$. The reason for this is that at the θ-point the polymer configuration forms part of a percolation cluster hull, and thus the presence of a virtual second polymer completing the hull is felt. This argument has been confirmed by simulations [8].

While the predicted values of C in the swollen phase and at the Manhattan lattice θ-point have been confirmed by simulations, there has been no confirmation of the predicted value $C = 24/7$ at the θ-point. There has also been some argument in the past whether the collapsed phase can indeed be identified with a dense phase. Moreover, the heuristic argument leading to a law-of-large-numbers could break down for the collapsed phase, as the chain is no longer swollen, so that a numerical verification of the predicted value $C = 4$ is especially important.

We note that there have also been simulations of interacting self-avoiding walks up to length $N = 300$ [9], suggesting that the results at the θ-point and in the collapsed phase are more consistent with a stretched exponential of the type $\exp(-|\theta|^\alpha/C\log N)$ with $\alpha \approx 1.5$. As a consequence, the scaling variable would change to $x = \theta/(\log N)^{1/\alpha}$.

Here we show, using simulations of interacting self-avoiding walks up to $N = 400$ with flatPERM, which allows simultaneous uniform sampling with respect to the number of interactions and the winding angle, that the winding angle distributions are indeed asymptotically Gaussian with a scaling variable $x = \theta/\sqrt{\log N}$, and that the variance grows linearly in $\log N$ with the value of C being in agreement with the theoretical predictions for swollen, critical, and dense polymers.

2. Simulation Details

A large ensemble of linear chain configurations was obtained on a square lattice by means of the flatPERM algorithm proposed by Prellberg and Krawczyk in [10], which is an extension of the Pruned and Enriched Rosenbluth Method (PERM) proposed in [11]. The algorithm is able to obtain statistical weights for a specific size and energy level (N, e). FlatPERM was successfully applied to several models in a lattice such as, for example, asymmetric interacting self-avoiding trails (ISAT) [12].

The flatPERM (multi-parametric flatPERM), described here, is able to obtain the aforementioned weights as a function of chain size, energy and any additional observable considered(N, e, O). An advantage of multi-parametric flatPERM is the estimation of micro-canonical partition function of any walker model as a function of N, e and the observable considered. In addition, a suitable observable considered can enhance the calculation pulling out "strange" configurations, or rather less statistical probable configurations.

A key variable in flatPERM is the ratio (r) between the the actual weight of a chain configuration of size n and energy m, $W_{n,m}$, and an estimation of it, $C^{est}_{n,m}$. If r is bigger than one we enrich, otherwise we prune. This ratio is changed accordingly in our version as

$$r = \frac{W_{n,m,k}}{C^{est}_{n,m,k}}, \tag{6}$$

where the k index corresponds to the value of the observable O in a configuration, and

$$C^{est}_{n,m,k} =< W >_{n,m,k}= \frac{1}{S}\sum_i W^{(i)}_{n,m,k}, \tag{7}$$

i.e, the average of all the generated samples i with size n, energy m and value of the observable k.

The observable implemented in this work is the winding angle (w) [3], i.e., the angle about a rod or reference point that is swept out by growing polymer or random walker. Winding angle is well defined in two dimensions and has been studied extensively in the literature for several kinds of walkers [5, 6, 8, 13, 14]. Winding angle is an interesting observable, since it gives geometric information about the degree of rolling up about the origin. Therefore, it is a excellent candidate to test our new implementation of flatPERM.

The flatPERM [10] algorithm has a technical problem due to the inaccuracy of the initial weights. This problem is even harder in our version, since the additional parameter can create degeneracy for each energy level m (weight). Degeneracy (d) produces a lot of initial enrichment, $r \gg 1$, because the estimated weight is a partial average of the initial configuration with same weight (energy). Then the initial run of the presented flatPERM version is characterized by a lot of enrichment, in order to obtain a proper estimation of the weight. This problem is particularly difficult if the observable is not very related with energy (weight), and it is a continuous function that, as in our case, should be discretized.

Histogram binning is a critical ingredient in this algorithm that will determine the degree of degeneracy, d. The smaller the binning the more enrichment that will be necessary. A priori an optimal binning will be one that reduces degeneracy, improves the generation of "less usual" walks, and ensures the corresponding histogram fits in the computer memory available. The calculations here presented have used a general histogram binning of 10 degrees. It is small enough to identify all the possible configurations for the smallest chains and the corresponding histogram fits in the computer memory available to us. However, this binning might be not optimal for big chains, given the big amount of winding angles possible.

Initial enrichment issues were solved integrating a certain percentage (g) of the degeneracy level around the location of the winding angle:

$$\hat{C}_{n,m,k} = \frac{1}{S} \sum_{i=k-a}^{k+a} \sum_{j} C_{n,m,i}^{(j)}, \tag{8}$$

where $a = d \cdot g/100$.

The advantages of this procedure are several: (i) it allows to get better initial estimations of the weights preventing too much enrichment; (ii) it can be equally applied to be energy to improve the issue above mentioned with initial wrong weights; (iii) it is size dependent and can be used with any kind of observable; (iv) the integration percentage, g, can be reduced along the run in order to increase the degeneracy (enrichment) and, therefore, the searching of "unusual" configurations.

On the other hand, this implementation suffers from two major disadvantages: (i) parameter dependency due to the binning; and (ii) is computationally expensive in terms of time and storage, given the large dimensional histogram that must be obtained. However, the second disadvantage can be easily solved with parallel computing.

Calculations here presented started with $g = 100$ % for 5×10^5 tours, that was subsequently reduced to 5 %. The total number of tours, S in eq 7 and 8, it is about 2×10^6. It is worth mentioning that our implementation allows restarting runs, saving the ful state of enrichment/prunning along the chain. Thus, it is possible combine several runs with different random seeds to decrease the correlations between samples mentioned in flatPERM [10].

3. Results

From the two-parameter histograms in the number of interactions and winding angle obtained, we can now extract winding angle distributions at several temperatures. We have chosen $\beta = 0$ for the swollen phase, $\beta = 0.8$ for the collapsed phase, and $\beta = 0.6673$ for the θ-point, using the best published estimate $\beta_\theta = 0.6673(5)$[15]. (We note in passing that previous work [9] used values 0, $\beta_\theta = 0.658(4)$, and 0.8.)

In Figure 1 we show winding angle distributions at infinite temperature ($\beta = 0$) for sizes $N = 100$ to $N = 400$, scaled such that the distributions have unit variance. Uniform sampling with respect to the winding angle enables us to sample the distribution over more than 100 orders of magnitude. We see that the central part of the distribution approaches a Gaussian distribution very well. Convergence in the tails is much slower, as is evident from the semi-logarithmic plot. However, this is not surprising, as the full distribution is supported on an interval that grows as $\sqrt{N}$ due to the largest winding angle being obtained by tightly wound spirals.

In Figure 2 we show winding angle distributions at the θ-point ($\beta = 0.6673$) for sizes $N = 100$ to $N = 400$, again scaled to unit variance. The central part of the distribution again shows a nice scaling collapse. Note the significant change of the probability in the tail of the distribution, due to the fact that configurations with large winding also have large energy and thus have significantly higher probability at lower temperatures.

In Figure 3 we show scaled winding angle distributions in the low temperature regime ($\beta = 0.8$) for sizes $N = 100$ to $N = 400$, again indicating scaling collapse.

Now that we have established that the central regions of the winding angle distributions collapse to the same distributions, we turn to the investigation of the growth of the winding angle variance in the different regimes. In Figure 4 we display the variance as a function of $\log N$. We also plot the three theoretical scaling predictions on these plots which are simply straight lines with slopes of the values of C being 2, 24/7 and 4 in the different regimes. There is clearly some curvature in the data coming from finite-size effects. However we can confidentially

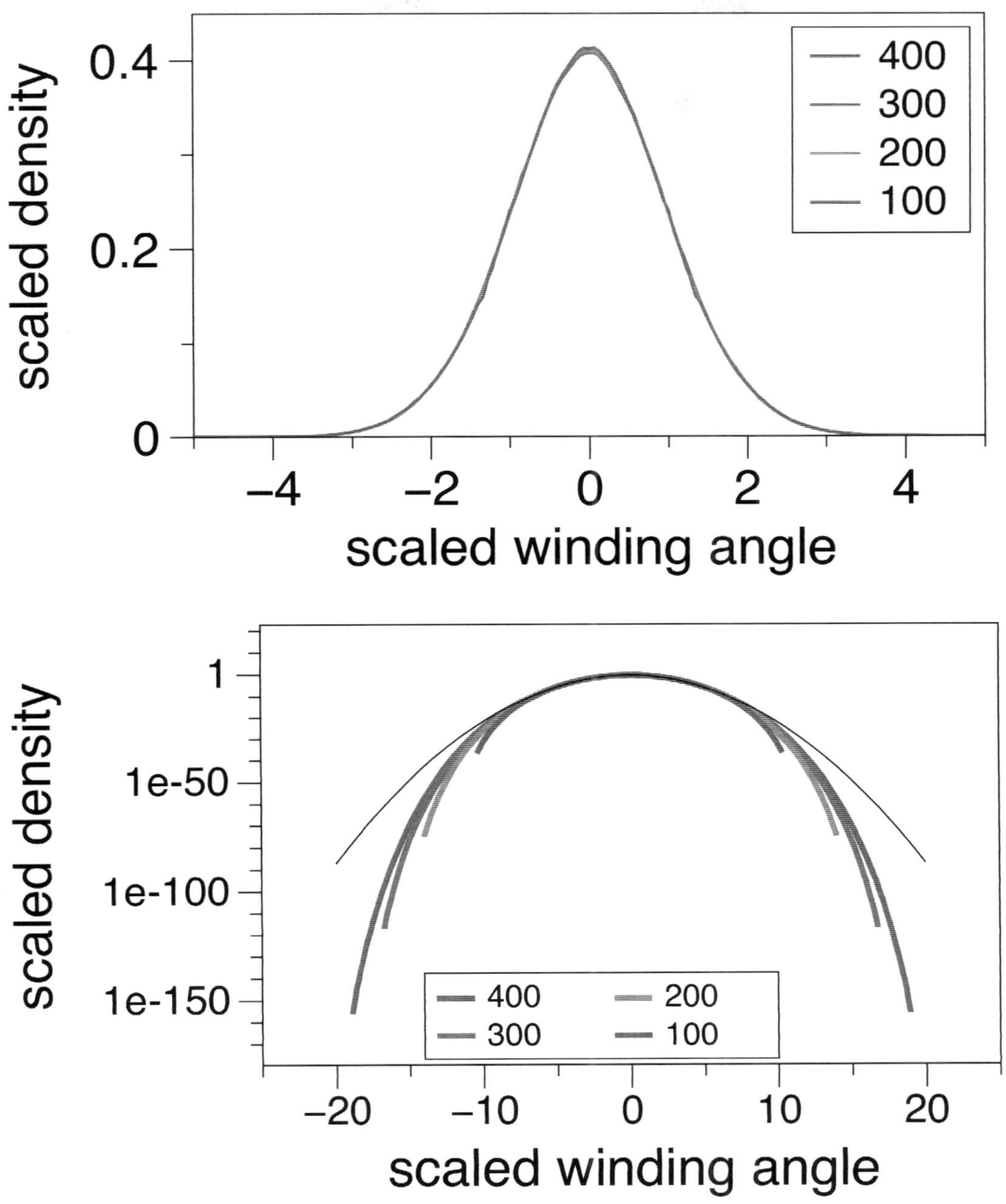

Figure 1. Top: normalised winding angle distributions at $N = 100$, 200, 300, and 400 for $\beta = 0$. Bottom: the same data in a semi-logarithmic plot, indicating the spread of the distribution as N increases. The thin line indicates the normal distribution $\exp(-x^2/2)/\sqrt{2\pi}$.

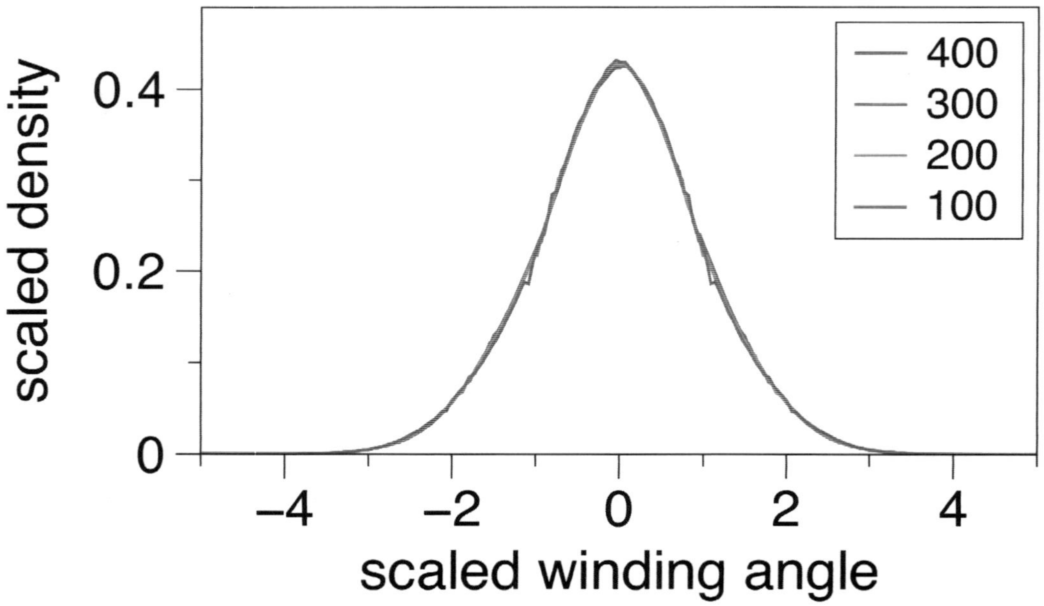

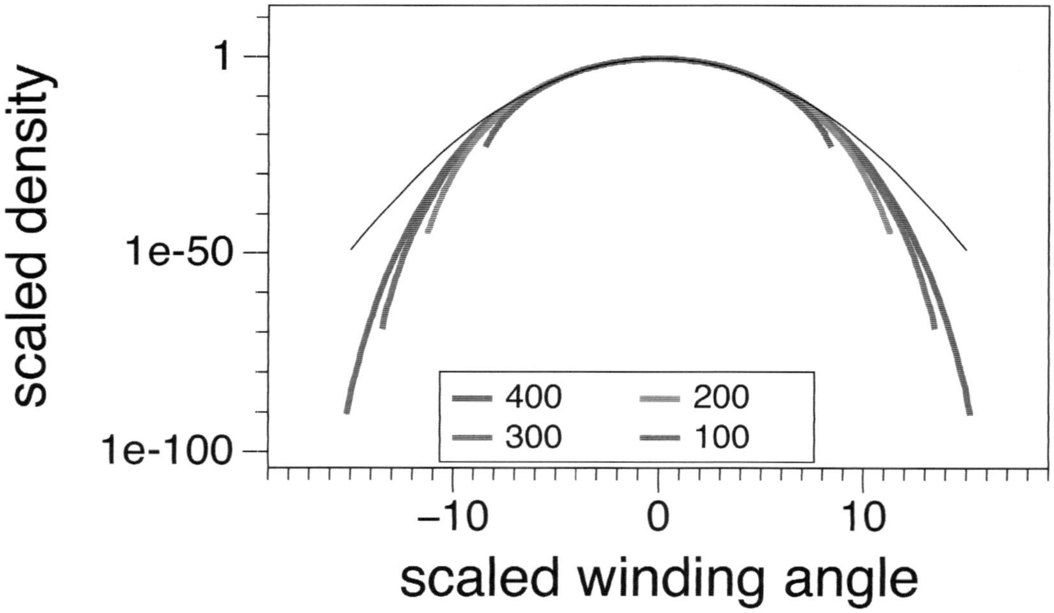

Figure 2. Top: normalised winding angle distributions at $N = 100$, 200, 300, and 400 for $\beta = 0.6673$. Bottom: the same data in a semi-logarithmic plot, indicating the spread of the distribution as N increases. The thin line indicates the normal distribution $\exp(-x^2/2)/\sqrt{2\pi}$.

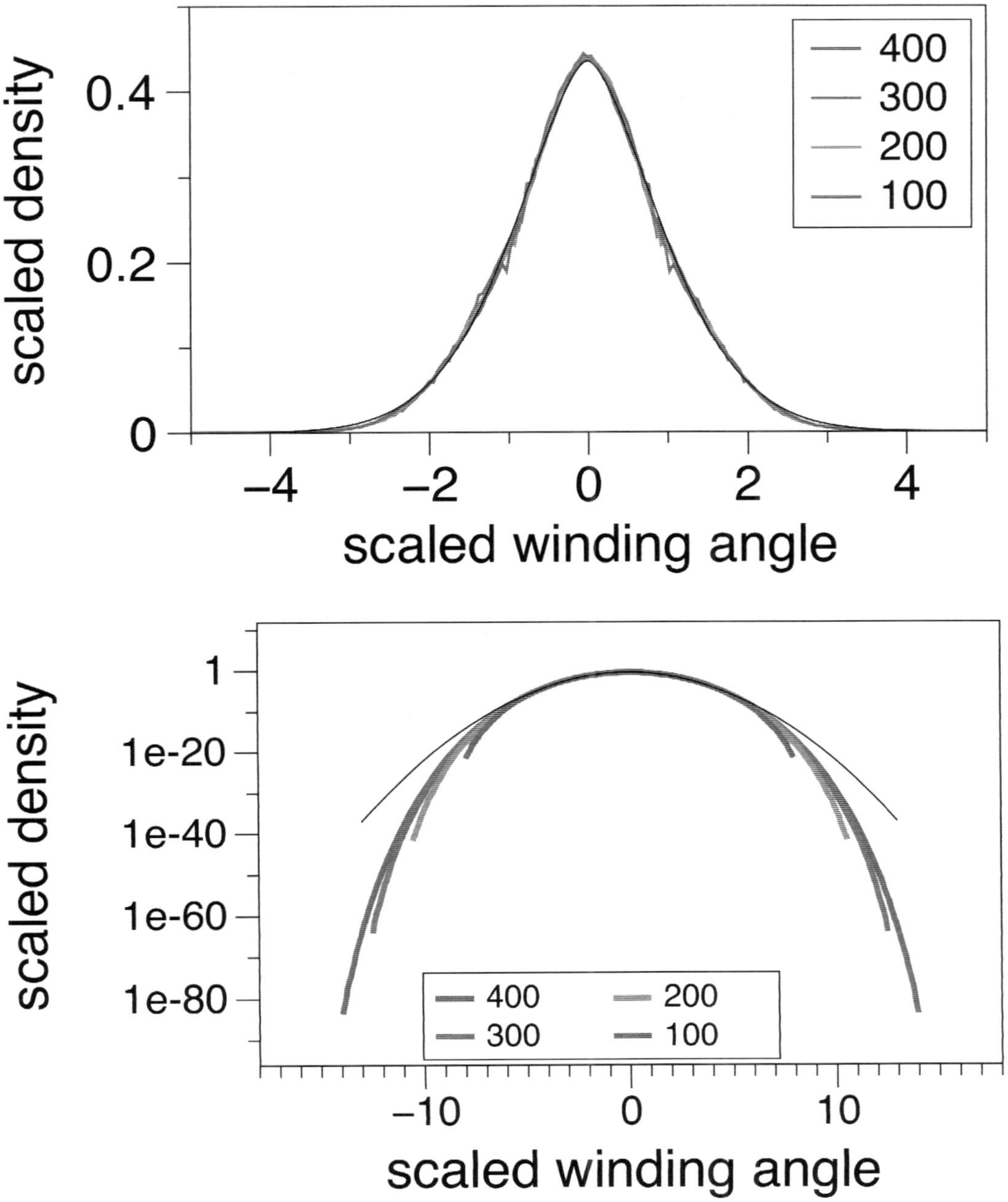

Figure 3. Top: normalised winding angle distributions at $N = 100$, 200, 300, and 400 for $\beta = 0.8$. Bottom: the same data in a semi-logarithmic plot, indicating the spread of the distribution as N increases. The thin line indicates the normal distribution $\exp(-x^2/2)/\sqrt{2\pi}$.

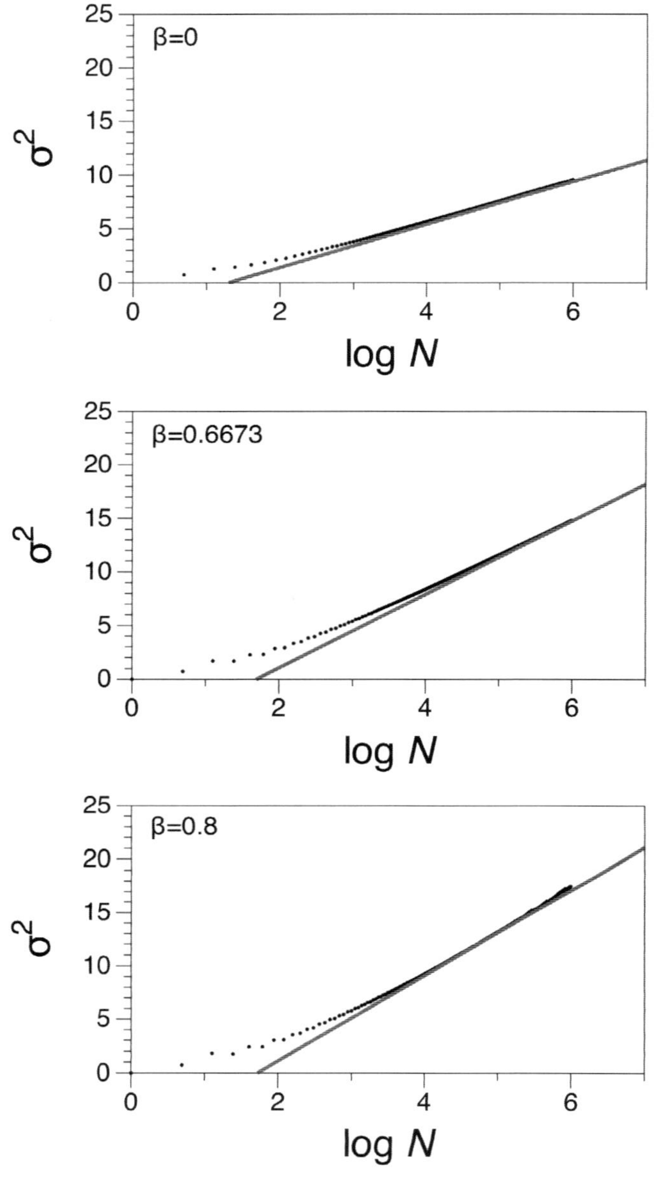

Figure 4. Shown is the variance of the winding angle distribution for lengths up to $N = 400$ (dots) at $\beta = 0$ (top), $\beta = 0.6673$ (middle), and $\beta = 0.8$ (bottom). The straight lines have slopes of 2 (top), 24/7 (middle), and 4 (bottom) illustrating the compatibility of the data with the theoretical scaling predictions.There is curvature at short lengths, as one would expect coming from corrections to scaling. There is also some noisiness and perhaps curvature in the data for the longest lengths in the low temperature figure (bottom) when compared to the theoretical straight line.

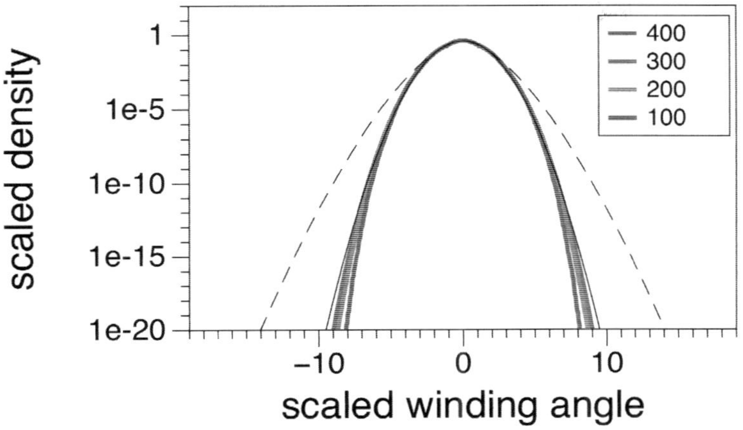

Figure 5. Normalised winding angle distributions at $N = 100, 200, 300$, and 400 for $\beta = 0.6673$. The thin line indicates the normal distribution $\exp(-x^2/2)/\sqrt{2\pi}$, whereas the dashed line indicates a stretched exponential with exponent $\alpha = 1.6$.

infer compatibility at both high temperatures ($\beta = 0$) and at the θ-point ($\beta = 0.6673$). At our low temperature value ($\beta = 0.8$) we make a more qualified statement as there are larger errors in our data and still some curvature at longer lengths, so state that the theoretical prediction of 4 for C is reasonable compatible with the data. In any case there is clear disagreement with the θ-point and low-temperature values of 2.75 and 3.14 given in [9].

Note also that the behaviour in the tails of the distribution does not indicate the presence of a stretched exponential. At the θ-point, Figure 5 shows that while it is entirely plausible that there is convergence to a normal distribution, convergence to a stretched exponential, which has much fatter tails, seems quite unlikely. The figure shows a stretched exponential with exponent $\alpha = 1.6$, somewhat larger even than the value of $\alpha = 1.54$ claimed in [9].

Acknowledgments
A L Owczarek thanks the School of Mathematical Sciences, Queen Mary, University of London for hospitality. T Prellberg acknowledges support by EPSRC grant EP/L026708/1.

References

[1] de Gennes P G 1979 *Scaling Concepts in Polymer Physics* (Ithaca: Cornell University Press)
[2] Spitzer F 1958 *Trans. Am. Math Soc.* **87** 187
[3] Rudnick J and Hu Y 1987 *J. Phys. A: Math. Gen.* **20** 4421
[4] Belisle C 1989 *Ann. Probab.* **17** 1377
[5] Saleur H 1994 *Phys. Rev. E.* **50** 1123
[6] Duplantier B and Saleur H 1988 *Phys. Rev. Lett.* **60** 2343
[7] Prellberg T and Drossel B 1998 *Physica A* **249** 337
[8] Prellberg T and Drossel B 1998 *Phys. Rev. E.* **57** 2045
[9] Chang I 2000 *Int. J. Mod. Phys. C* **11** 721
[10] Prellberg T and Krawczyk J 2004 *Phys. Rev. Lett.* **92** 120602
[11] Grassberger P 1997 *Phys. Rev. E* **56** 3682
[12] Bedini A, Owczarek A L and Prellberg T 2013 *Phys. Rev. E* **87** 012142 (8pp)
[13] Samokhin K 1998 *J. Phys. A: Math. Gen.* **31** 8789
[14] Rudnick J and Hu Y 1988 *Phys. Rev. Lett.* **60** 712
[15] Caracciolo S, Gherardi M, Papinutto M and Pelissetto A 2011 *.Phys. A: Math.Theor.* **44** 115004

Information processing occurs via critical avalanches in a model of the primary visual cortex

G. S. Bortolotto[1], M. Girardi-Schappo[1], J. J. Gonsalves[1], L. T. Pinto[2], M. H. R. Tragtenberg[1]

[1] Departamento de Física, Universidade Federal de Santa Catarina, 88040-900, Florianópolis, Santa Catarina, Brazil

[2] Departamento de Engenharia Química e de Alimentos, Universidade Federal de Santa Catarina, 88040-900, Florianópolis, Santa Catarina, Brazil

E-mail: marcelotragtenberg@gmail.com

Abstract. We study a new biologically motivated model for the Macaque monkey primary visual cortex which presents power-law avalanches after a visual stimulus. The signal propagates through all the layers of the model via avalanches that depend on network structure and synaptic parameter. We identify four different avalanche profiles as a function of the excitatory postsynaptic potential. The avalanches follow a size-duration scaling relation and present critical exponents that match experiments. The structure of the network gives rise to a regime of two characteristic spatial scales, one of which vanishes in the thermodynamic limit.

1. Introduction

Brain criticality has gained wide attention in the last years [1, 2, 3]. Several studies have shown how multiple cognition features are improved when considering the brain as a critical system. Some of these features are the optimization of response dynamic range [4, 5] and the enhancement of memory and learning processes [6, 7], computational power [8, 9], information processing flexibility [10] and network processing time [11].

The visual system has such a collection of experimental results [12, 13] that makes it suitable for validating models. Recently, some authors probed cortical tissues and sometimes specifically the visual cortex looking for neuronal avalanches both *in vitro* and *in vivo* [14, 15, 16, 17]. We developed a model of the primary visual cortex (V1) based on experimental constraints in order to study how do the microscopic details of the system shape the avalanche dynamics and statistics.

The stimulation of the retina of our model gives rise to avalanches. The avalanches spread radially inside each layer and through a branching process across different columns. The columnar structure and the whole network excitation are evident in the avalanches size and duration distributions which present two distinct power-law (PL) behaviors for a given *excitatory postsynaptic potential* (EPSP) range. We computed the exponents of these distributions and checked that they obey Sethna's scaling law [18] even when the system is adjusted outside of the critical regime.

Next section is dedicated to describe the model dynamics and structure. Section 3 brings results and discussions concerning the information processing of the network via avalanche

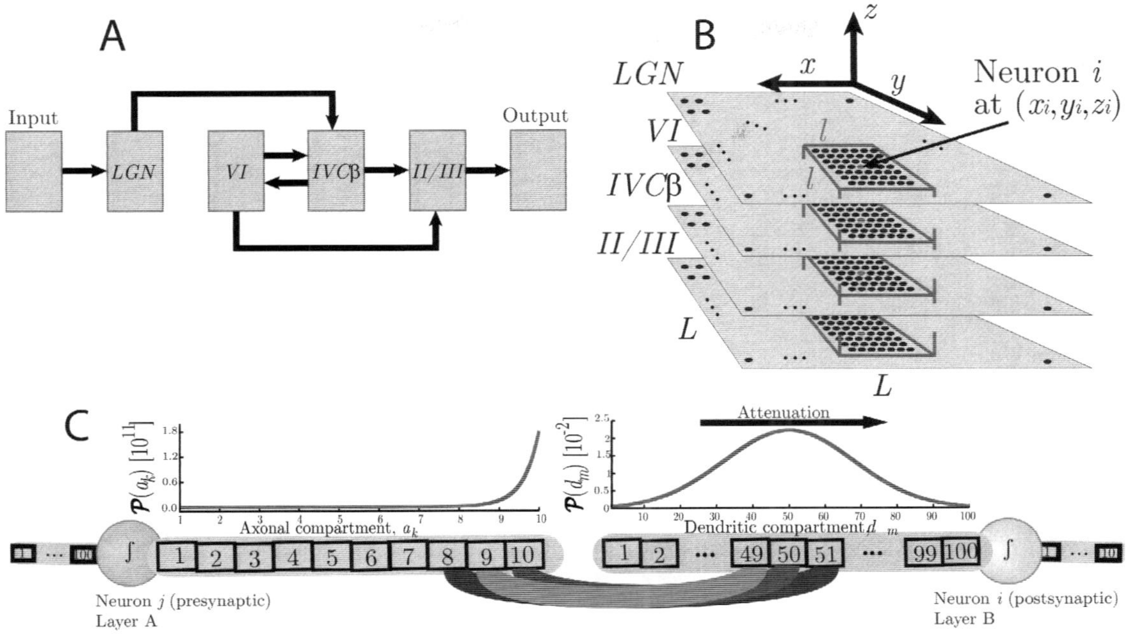

Figure 1. The V1 model. A: Architecture of the network. B: Spatial organization of the network of $N = 4L^2$ neurons. Each neuron in the network is at the center of a column of size $N_c = 4l^2 = 196$ neurons defined by the algorithm used to create synapses. This columnar structure in highlighted in red in Panel B. C: Compartmental scheme of neurons. Neuronal synapses start preferentially from the end of the axon (left), as given by the probability $\mathcal{P}(a_k)$ of choosing a presynaptic axonal compartment a_k. The dendritic postsynaptic compartment, d_m, is chosen with a Gaussian probability $\mathcal{P}(d_m)$ with mean 50 and standard deviation 10. Most synapses lay in the middle of the dendrite (right).

activity. After that, we conclude reviewing the main results and we point some possible extensions for this work.

2. Model

The model developed by Andreazza & Pinto is composed of six square layers [19]. The layers are connected to each other following a feed-forward mechanism with a single loop presented in Fig. 1A. No lateral connections are present inside the layers. The signal propagates from the retina (Input layer) to the secondary visual cortex V2 (Output layer).

The four internal layers have linear size L and correspond to the form recognition pathway within V1 (composed of layers II/III, IVCβ and VI) plus the lateral geniculate nucleus (LGN). The LGN layer consists of only its parvocellular neurons. The synapses of such neurons are mostly connected to V1 layer IVCβ [20, 21, 22, 23].

2.1. Network structure

The network has a total of $N_{full} = 2N_{io} + N$ elements. The four internal layers (LGN, VI, IVCβ and II/III) are composed by L^2 neurons each, summing up to a total of $N = 4L^2$ neurons. The neurons are composed of compartmental dendrites and axons connected by a soma compartment as seen in Fig. 1C. Each of the input/output layers has a total of $N_{io} = (10L)^2$ elements.

Table 1. The quantity of attempted synapses per presynaptic element from a specific layer (rows) to another layer (columns). Presynaptic element may be a photoreceptor (from Input layer) or a neuron (from internal layers). These values are predefined in the beginning of the simulation, but are not exactly achieved in practice because of free boundary conditions.

From \ To	LGN	VI	IVCβ	II/III	Output
Input	1	-	-	-	-
LGN	-	-	500	-	-
VI	-	-	1100	350	-
IVCβ	-	600	-	700	-
II/III	-	-	-	-	100

The Input layer represents the photoreceptors of the retina. The Output layer consists of axonal terminals which connect to dendrites in V2. Each neuron of the LGN is placed in front of a matrix of 100 photoreceptors of the Input layer. A similar structure is present on the other end of the network where each neuron in layer II/III randomly sends 100 synapses towards a 10×10 matrix in V2.

Layers are positioned parallel to the $x - y$ plane and are stacked along the z direction (see Fig. 1B). Each presynaptic neuron j has a spatial position $\vec{r}_j = (x_j, y_j, z_j)$. The number of synapses that each neuron of each layer sends towards its adjacent layers are fixed according to Table 1.

Every postsynaptic neuron is chosen in the adjacent layer with the use of a bidimensional Gaussian distribution, $\mathcal{P}_G(x, y; x_j, y_j, \sigma_c)$ centered in the presynaptic neuron (x_j, y_j) position with standard deviation $\sigma_c = 3$ inside a limited region of $l^2 = 7 \times 7$ neurons. This mechanism creates a columnar structure as shown in Fig. 1B.

Each column has approximately $N_c = 4l^2 = 196$ neurons. When taking into account the loop between layers IVCβ and VI, the columns can be considered to have $N_c = 5l^2 = 245$ neurons. Such structure is very important for the simulation results. Synapses sent by neurons on the border of each layer that fall outside of the adjacent layer are simply ignored (free boundary conditions).

Once postsynaptic neurons are chosen for every presynaptic neuron, we pick one axonal compartment $a_k^{(j)}$ of the presynaptic neuron j with exponential probability $\mathcal{P}_E(k) = (10/4)\exp(10k/4)$ and one dendritic compartment $d_m^{(i)}$ of the postsynaptic cell i, chosen with Gaussian probability $\mathcal{P}_G(m; d^{(c)}, \sigma_d)$ to form each synapse. Here $d^{(c)} = 50$ and $\sigma_d = 10$. These distributions are plotted in Fig. 1C.

Synaptic distribution over dendritic compartments is chosen such that the probability of receiving a synapse in the dendrite is maximum at its center [24] and that any consecutive synapses are uncorrelated. Axonal compartments distribution is chosen such that most of the synapses come out of the axon's end, making the signal travel as far as possible.

2.2. Neuronal and synaptic dynamics

Each neuron i is composed of a compartmental dendrite $[d_m^{(i)}(t);\ m = 1, \cdots, 100$, Eq. (1)], the soma $[v_i(t)$, Eq. (2)], and a compartmental axon $[a_k^{(i)}(t);\ k = 1, \cdots, 10$, Eq. (3)]. The action potential advances one compartment per time step t. This potential comes from the dendrites, passes through the soma and finally reaches the last axonal compartment:

$$\begin{aligned} d_1^{(i)}(t+1) &= \lambda E \sum_{j,n} a_n^{(j)}(t) \;, \\ d_k^{(i)}(t+1) &= \lambda \left[d_{k-1}^{(i)}(t) + E \sum_{j,n} a_n^{(j)}(t) \right], \text{for } k > 1, \end{aligned} \tag{1}$$

$$v_i(t+1) = \begin{cases} \Theta\left(d_{100}^{(i)}(t) - v_T\right) & , \text{ if } v_i(t) = 0 \;, \\ -R & , \text{ if } v_i(t) = 1 \;, \\ v_i(t) + 1 & , \text{ if } v_i(t) < 0 \;, \end{cases} \tag{2}$$

$$\begin{aligned} a_1^{(i)}(t+1) &= \Theta\left(v_i(t)\right) \;, \\ a_k^{(i)}(t+1) &= a_{k-1}^{(i)}(t), \text{for } k > 1, \end{aligned} \tag{3}$$

where $\lambda = 0.996$ is the attenuation constant (following experimental results [24]), $v_T = 10$ mV is the firing threshold, R is the refractory period (which is sufficient to avoid self-sustained activity within the interlayer loop) and $E > 0$ is the EPSP (of the order of 1 mV [24, 25, 26]). The double sum in Eq. (1) is over all the axonal compartments of presynaptic neuron j connected to the dendritic compartment of the postsynaptic neuron.

The different amount of dendritic and axonal compartments for each neuron takes into account the different velocities of signal propagation for a time step of 1 μs. In addition, the soma time step is of 1 ms, so that avalanches have a larger time scale than the propagation of the signal through the neurons compartments. The travelled distance in the dendrites is about 100 μm and for the axons 1000 μm [27]. Initial conditions are $a_k^{(i)}(0) = d_m^{(i)}(0) = v_i(0) = 0 \;\forall\; (k, m, i)$ for all the neurons of the internal layers. A square of 30×30 photoreceptors is flashed in the first time step in order to spark activity. Results are stable for different locations of the retina activation stimulus.

3. Results and discussion

We performed about one hundred simulations for each network linear size, L, and each EPSP, E. The activity of the network is the sum of all the neurons firings at each time step, $A(t) = \sum_{i=1}^{N} \delta_{v_i(t),1}$, where $\delta_{a,b}$ is the Kronecker delta. An avalanche size is defined as the total activity between two consecutive instants of inactivity (similarly to the experimental procedure [15, 28]), and the avalanche duration is simply the time interval between these consecutive instants of inactivity,

$$s(n+1) = \sum_{t=t_n}^{t_{n+1}} A(t) \;, \tag{4}$$

$$T(n+1) = t_{n+1} - t_n \;, \tag{5}$$

where $A(t_n) = A(t_{n+1}) = 0$.

On the critical point, the distributions of s and T are assumed to scale as [29]

$$\mathcal{P}(s) \sim s^{-\alpha} \mathcal{G}_s\left(s/s_c\right) \;, \tag{6}$$

$$\mathcal{P}(T) \sim T^{-\tau} \mathcal{G}_T\left(T/T_c\right) \;, \tag{7}$$

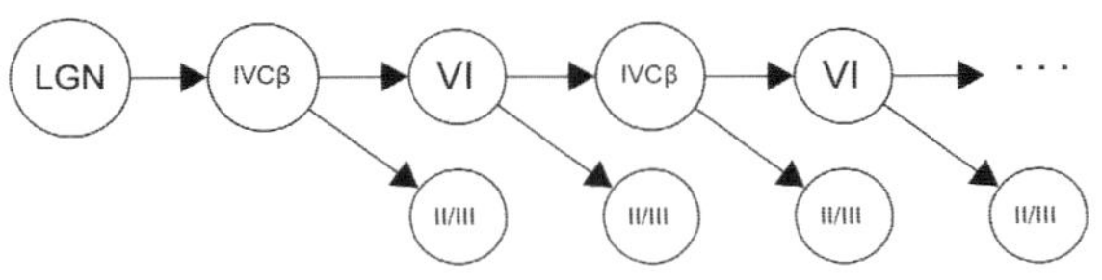

Figure 2. Schematic representation of signal branching process in the network. This follows from the network architecture in Fig. 1.

where s_c and T_c are the cutoff values of s and T, $\mathcal{G}_{s,T}(x)$ are scaling functions that describe how the cutoff of the distributions scale with system size and α and τ are scaling exponents that should follow Sethna scaling relation [18]:

$$a = \frac{\tau - 1}{\alpha - 1} , \tag{8}$$

where a is a growth exponent relating avalanches size and duration,

$$\langle s \rangle \sim T^a . \tag{9}$$

The scaling law of Eq. (8) must be obeyed if Eqs. (6) and (7) hold within a large enough interval of s and T.

3.1. Spatio-temporal profile of avalanches

Activity is initiated in the LGN and sent towards layer IVCβ inside a small localized region due to the columnar structure of the network. This small activated region in layer IVCβ acts as a seed for firing localized activity in the adjacent layers, VI and II/III. The activity of layer II/III is just sent over to V2 while activity from layer VI serves as another seed for firing localized activity again in layer IVCβ. Now, the neurons that have already fired in layer IVCβ are refractory, so only the neurons around the first seed region will fire this time. This process constitutes a branching process between layers, as illustrated in Fig. 2 and in Fig. 3 top left panel.

As a consequence of the branching process, the activity spreads as spiking circular waves inside each layer. The spatial profile of these circular avalanches is in Fig. 3. The top left panel illustrates how the activity spreads throughout the network internal layers starting in the LGN. Red and white colors indicate the sequence of time steps, although not necessarily immediate consecutive time steps. Activities represented in red always come after activities represented in white and *vice-versa.*

Two features should be noticed with the help of Fig. 3: firstly it shows that there is a qualitative change in the behavior of the network as E increases: for $E < 1.12$ mV the signal does not reach the border of the layers. For $1.12 \leq E \leq 1.19$ mV, there is a vanishing probability of reaching the borders whereas for $E > 1.19$ mV the activity always reaches the border. This behavior indicates that $E = 1.19$ mV is a critical point.

Secondly, notice that as E increases, layer II/III activity starts to happen simultaneously with activity in both layers VI and IVCβ because less and less presynaptic neurons (in the seed region) are needed to cause any neuron to fire. Such profile reduces the silence intervals between consecutive avalanches (see Fig. 4A). Thus for large $E = 13$ mV, the activity spatial profile reduces to a two-dimensional radial wave front propagating simultaneously in the three layers via a single very large avalanche.

Columns therefore play a significant role by propagating activity only for $E \geq 1.12$ mV but yet with E sufficiently small so that a neuron needs a considerable amount of presynaptic active neurons to pass on the signal forward.

The temporal profile of avalanches is shown in Fig. 4A for four typical EPSP. Notice that there is also a qualitative change in the activity temporal profile as E increases. The network processing time (the time interval between stimulation and the last spike in the network) continuously increases for increasing $E < 1.21$ mV. However, the variance of the processing time is maximal for the whole interval $1.12 \leq E \leq 1.19$ mV. The maximum variance for a finite interval indicates that the system is very flexible to process information inside this region and also might signalize the presence of a critical Griffiths phase [30, 31].

Processing time is maximum for $E = 1.21$ mV and decreases asymptotically for increasing $E > 1.2$ mV because the activity across layers becomes entangled. The entanglement of activity

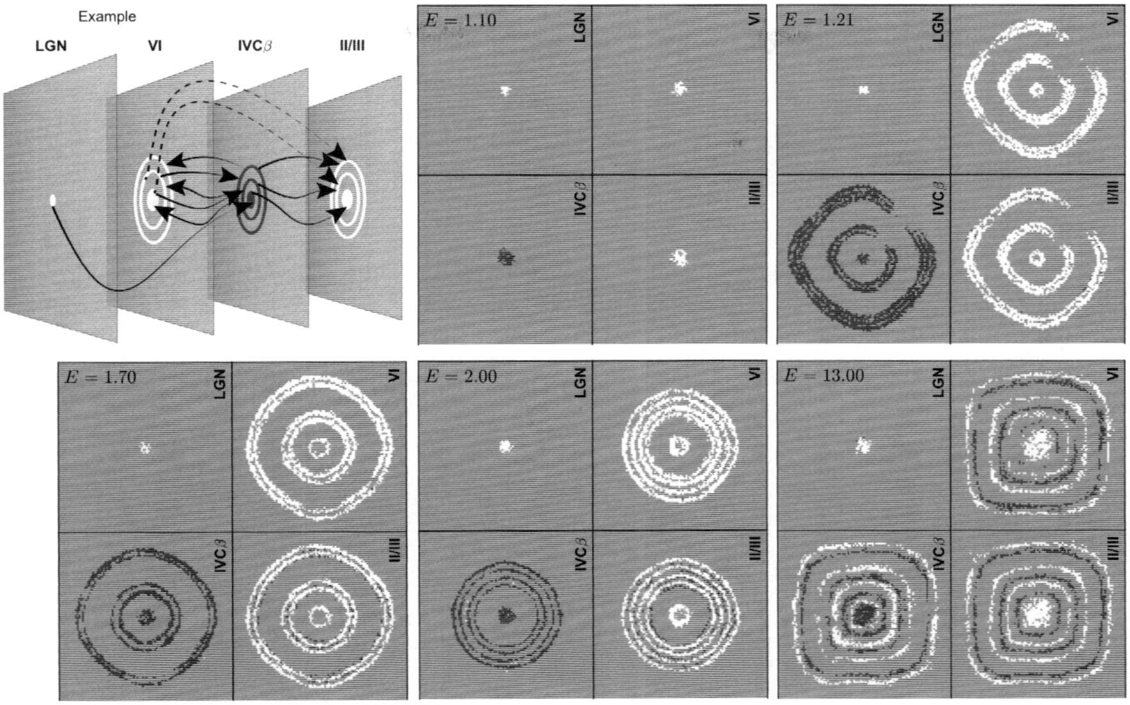

Figure 3. Spatial profile of avalanches in the internal layers for different E. Each circle is a snapshot of the network in a different time step. Top left panel illustrates the temporal order of events which is useful to analyze the other panels: Activity starts in the center of the LGN and then follows the directions pointed by the arrows. Circle colors are used to indicate temporal sequence: activity in red always follows activity in white and so on, although not in the immediate next time step. $E = 1.10$ mV: the signal travels through all layers but is unable to reach the border of the layers; layer II/III spikes only together with layer VI. In fact, this means that avalanches are spreading only inside columns in a sort of branching process (see Fig. 2). Starting from $E = 1.12$ mV, activity has probability to reach the border of the layers. For $E = 1.21$ mV, the activity always reaches the border; layer II/III spikes together with VI, although some of II/III neurons spike together with IVCβ due the signal coming from VI. The panels with $E = 1.70$ mV, 2.00 mV show how the entanglement of activity in layers IVCβ and II/III strengthens for larger EPSP; the avalanches become larger due to less inactivity intervals (see Fig. 4A). For $E = 13.00$ mV activity in layer VI, IVCβ and II/III are so entangled that a single large avalanche appears and radially spreads inside the network.

decreases the waiting times between avalanches. Such behavior is evident for $E = 2.0$ mV, in which there is a large avalanche ranging from about $t \approx 900$ until $t \approx 2100$ time steps, and also for $E = 13$ mV.

3.2. Avalanche distributions

We computed the normalized probability and complementary cumulative distributions of the avalanche sizes and duration. The complementary cumulative distribution (henceforth referred only as cumulative distribution) is the probability of measuring an avalanche of size s or greater.

In our model the information processing may have four different avalanche profiles (see Fig. 4B). First, for $E < 1.12$ mV only small avalanches occur. The signal is still able to travel throughout all the layers, but only a few neurons are in fact activated – this is the *disordered*

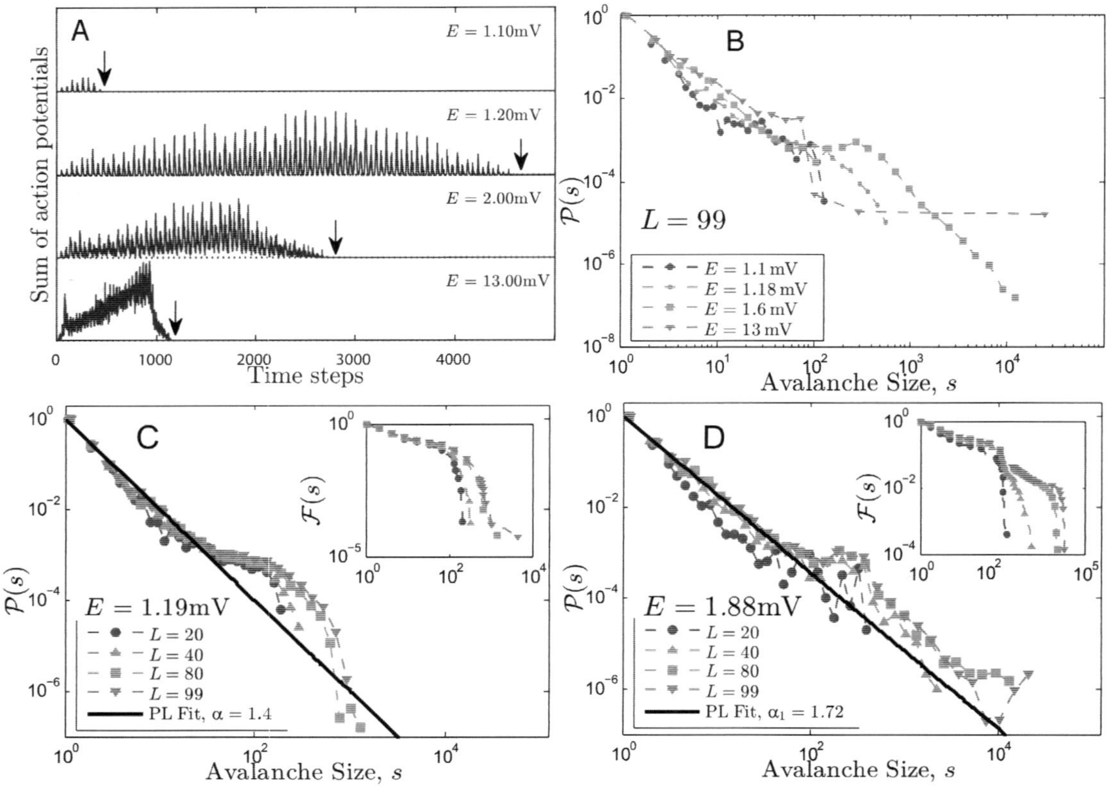

Figure 4. A: Temporal profile of the avalanches for different E. Arrows mark total processing time; B: Distributions $\mathcal{P}(s)$ of avalanches sizes for four typical E values, one in each regime described in the text. An avalanche size is the total activity between two consecutive instants of inactivity in Panel A; C: $\mathcal{P}(s)$ for multiple network sizes L on the critical point $E = 1.19$ mV (notice how the cutoff scales with system size); D: $\mathcal{P}(s)$ for different L on the weakly ordered regime for $E = 1.88$ mV (notice the bump in the middle of the distribution localized around $s = N_c \approx 200$ neurons); The insets in panels C and D show the corresponding complementary cumulative distributions $\mathcal{F}(s)$ for each L and E. Dashed lines are only present to guide the eyes.

regime (or subcritical phase). Second, the region for $1.12 \leq E \leq 1.19$ mV is *critical*: there are many small and large avalanches and each layer activity is disentangled from the other, as discussed in Section 3.1 and shown in Fig. 3. Fig 4C shows the PL fit of the avalanche size distribution on the critical point. Such distributions equal all the distributions in the critical E range. Notice how the largest avalanche, and hence s_c and T_c, scale with system size.

The third behavior lies in the range $1.20 \leq E \leq 2.00$ mV, where large avalanches begin to dominate network activity, making the distributions tails heavier as E is increased. Fig 4D shows the distribution for $E = 1.88$ mV and many system sizes. In this EPSP range, we can identify two different PL behaviors, one for small avalanches and another for large ones due to the entangling of activity across different layers. The characteristic bump in the middle of the distribution happens exactly for avalanches of size $s = N_c \approx 200$. Thus, small avalanches happen mostly because of disentangled activity inside the network columns whereas large avalanches are due to activity that span multiple layers. We call this regime *weakly ordered.*

Finally, for $E > 2.00$ mV the network is swept by one large avalanche (see Fig. 4A and

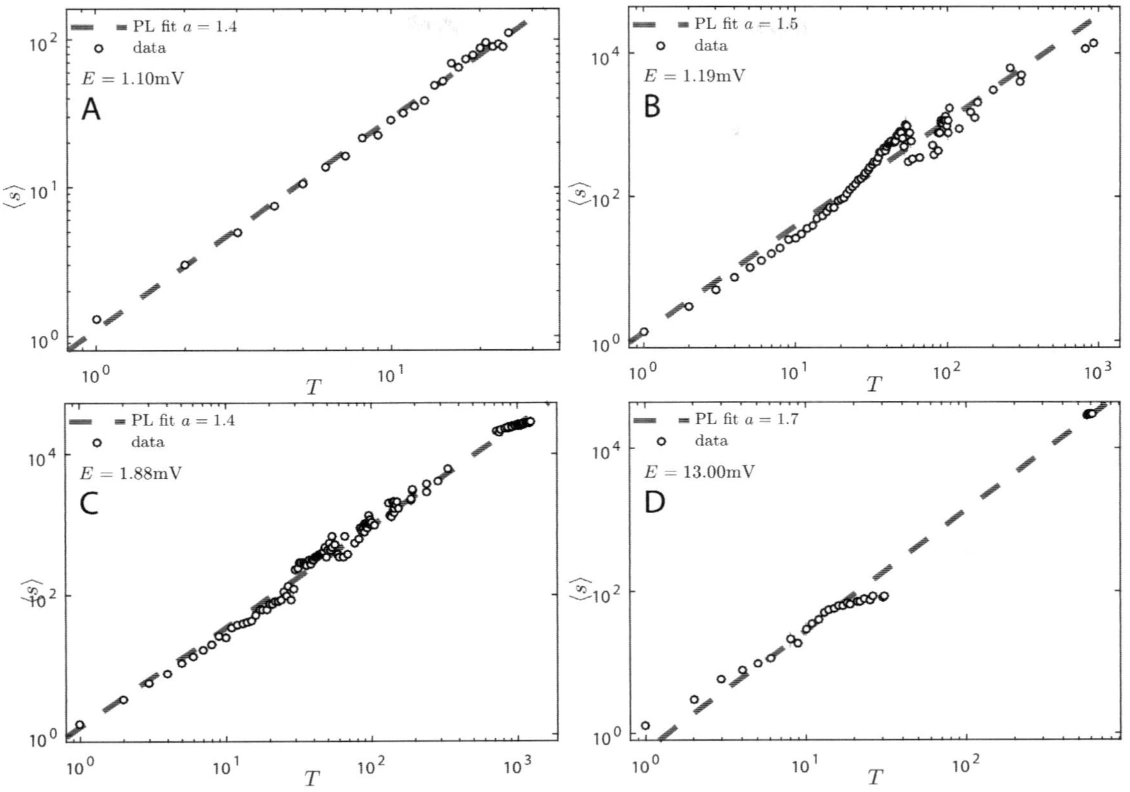

Figure 5. The size-duration scaling relationship for the avalanches for different E. The points are the simulation data and the dashed lines are the corresponding fits for $\langle s \rangle \sim T^a$ yielding a_{Fit} in Table 2. A: $E = 1.1$ mV, only very small avalanches are present; B: $E = 1.19$ mV is the critical point and presents PL distributed avalanches for all scales; C: $E = 1.88$ mV is in the weakly ordered regime and presents a small gap between very large avalanches and the remaining ones (each large avalanche corresponds to a new realization of the simulation); D: for $E = 13$ mV there is a huge gap between the many very large avalanches and some remaining small avalanches, where each large avalanche is a simulation realization.

Fig. 4B) that spans through all the layers simultaneously during the entire simulation time. The small avalanches regime is still a PL (purple triangles in Fig 4B is a PL for $s \lesssim 100$), although these avalanches are just the spark in the LGN used to start activity. This behavior is stable for any initial condition, so this is the *strongly ordered* regime. Together, weakly and strongly ordered regimes make up a supercritical phase.

We calculated the critical exponent α [Eq. (6)] in Figs. 4 B, C and D with the collapse of the cumulative distributions for different system sizes, L. This calculation is consistent with a Maximum Likelihood PL fitting to the PL regime of the distributions [11]. The same procedure was applied to calculate exponent τ [Eq. (7)]. Some values for α and τ are presented in Table 2 (second and third columns). These values are close to the ones obtained in experiments [14, 17] and are close to avalanche exponents of absorbing state phase transitions [32]. The collapse of avalanche distributions also yield dynamical spreading exponents that describe how s_c and T_c increase with L. The growth exponent of the largest avalanche is calculated elsewhere along with a rigorous study of the discussed phase transition [11].

We find that Sethna's relation holds even outside of the critical regime, contrary to commonly

Table 2. The duration and size power-law distribution exponents are shown in columns τ and α, respectively. Column a_{Dist} is obtained with the use of Eq. (8). The values of a_{Fit} are obtained by fitting simulation data in Fig. 5. The percentual error between these two approaches is shown in the last column. τ, α and a_{Fit} have an associated error of 5% each.

E (mV)	τ	α	a_{Dist}	a_{Fit}	Error (%)
1.10	1.788	1.602	1.309	1.426	-8.17
1.19	1.596	1.390	1.531	1.470	4.19
1.88	1.737	1.491	1.500	1.434	4.58
13.0	1.520	1.350	1.483	1.688	-12.1

believed [17]. We calculate Sethna exponent a for the four regimes of E using Eq. (8) using the values obtained for τ and α yielding a_{Dist}, the fourth column in Table 2. We also fit the curve Eq. (9) to data in Fig. 5, yielding a_{Fit}, the fifth column in the same table. The percentual error between these two values for a is presented in column six. The errors for a_{Dist} are inherited from the calculation of τ and α. These scaling exponents are obtained with a fit on the PL regions of the avalanches size and duration distributions, each having associated errors of 5%. Also, the values of a_{Fit} have associated error of 5%. We can then consider the relative errors between the two quantities to be negligible. The cutoff s_c for both distributions is small, of the order $L^{D'}$, with $D' = 1.1$ [11], and this reflects on the fairly high error values.

4. Conclusion

We studied a model presenting several dynamical and structural features of the primary visual cortex. Some of which have been proved to be of fundamental importance in the avalanche dynamics of the system: the columnar structure and the extended body of the neurons. The first is needed for the signal to activate the entire network even when the excitatory postsynaptic potential is not strong enough to have a one-to-one neuronal activation. The latter generates the intrinsic avalanche dynamics of the system since the interval between avalanches is a consequence of the propagation of action potentials in dendritic and axonal compartments.

We identified four regimes of activity in the system: a disordered regime, in which avalanches are not able to reach system lateral boundaries and thus avalanches also do not scale with system size (nevertheless small avalanches are PL distributed); a critical regime, in which avalanches are PL distributed and avalanche cutoffs scale with system size; a weakly ordered regime, in which avalanches are PL distributed and are processed in two characteristic size and time scales – one corresponding to the amount of neurons in a column and the other to the system boundary; and finally a strongly ordered regime, in which there is a single large dominating avalanche which scales with system size, although the spark avalanches that initiate activity are also PL distributed.

We also verified that Sethna's relation holds for the four regimes and thus we claim that having PL distributed avalanches (both in size with exponent α and in duration with exponent τ) that follow the scaling relation $a = (\tau - 1)/(\alpha - 1)$ is not sufficient to tell whether the system is in the critical state. Instead, a good criterion is to have PL avalanches with cutoff that scale with system size. Nevertheless, scaling of the cutoff is only a weak criterion to determine criticality. The strong criterion would be to define an order parameter and its associated susceptibility and then check for the divergence of susceptibility [11].

Our next steps will be to modify some aspects of the model, e.g. changing the form of the input signal in order to to see if the behavior and relationships presented by the network are maintained; adding heterogeneity in E parameter in order to reproduce its real distribution in the cortex [25, 26]; adapting the excitatory field of each neuron in deeper layers; using synaptic dynamics or plasticity [17, 33, 34] to model adaptability [35]; and adding lateral inhibition.

Moreover, we hope to provide here a kinematic framework for microscopic cortical modeling. We conjecture that our model pertains to Dynamical Percolation-like universality class because its dynamics resembles that of the generalized epidemic process with immunization [36, 37], which is part of a broad universality class of absorbing state phase transitions [32].

Acknowledgments

We thank the organizers of the VIII BMSP for the opportunity to show some of the results obtained in this work. Authors GSB and MGS thank the financial support of FAPESC and CNPq agencies, respectively. We also thank S. Boettcher for valuable discussions.

References

[1] Chialvo D R 2010 *Nat. Phys.* **6** 744–750
[2] Plenz D and Niebur E (eds) 2014 *Criticality in Neural Systems* (Weinheim, Germany: Wiley-VCH) ISBN 978-3-527-41104-7
[3] Beggs J 2015 *Phys. Rev. Lett.* **114(22)** 220001
[4] Kinouchi O and Copelli M 2006 *Nat. Phys.* **2** 348–351
[5] Shew W L, Yang H, Petermann T, Roy R and Plenz D 2009 *J. Neurosci.* **29(49)** 15595–15600
[6] de Arcangelis L, Perrone-Capano C and Herrmann H J 2006 *Phys. Rev. Lett.* **96** 028107
[7] Scarpetta S and de Candia A 2013 *PLoS One* **8(6)** e64162
[8] Shew W L and Plenz D 2013 *Neuroscientist* **19(1)** 88–100
[9] de Arcangelis L and Herrmann H J 2010 *Proc. Natl. Acad. Sci. USA* **107(9)** 3977–3981
[10] Mosqueiro T S and Maia L P 2013 *Phys. Rev. E* **88(1)** 012712
[11] Girardi-Schappo M, Bortolotto G S, Gonsalves J J, Pinto L T and Tragtenberg M H R 2015 *arXiv:1509.08954 [q-bio.NC]*
[12] Albright T D, Jessell T M, Kandel E R and Posner M I 2000 *Neuron* **25** S1–S55
[13] Clifford C W, Webster M A, Stanley G B, Stocker A A, Kohn A, Sharpee T O and Schwartz O 2007 *Vision Res.* **47(25)** 3125–3131
[14] Beggs J M and Plenz D 2003 *J. Neurosci.* **23(35)** 11167–11177
[15] Ribeiro T L, Copelli M, Caixeta F, Belchior H, Chialvo D R, Nicolelis M A L and Ribeiro S 2010 *PLoS ONE* **5(11)** e14129
[16] Hahn G, Petermann T, Havenith M N, Yu S, Singer W, Plenz D and Nikolić D 2010 *Journal of Neurophysiology* **104(6)** 3312–3322
[17] Shew W L, Clawson W P, Pobst J, Karimipanah Y, Wright N C and Wessel R 2015 *Nat. Phys.* **11** 659–663
[18] Sethna J P, Dahmen K A and Myers C R 2001 *Nature* **410** 242–250
[19] Andreazza J K and Pinto L T 2006 *Prooceedings of 2nd LNCC Meeting on Computational Modelling* (Petrópolis, RJ, Brazil: LNCC)
[20] O'Kusky J and Colonnier M 1982 *J. Comp. Neurol.* **210**(3) 278–290
[21] Lund J S 1984 *The Cerebral Cortex* vol 1 ed Peters A and Jones E G (New York, USA: Plenum Press) chap 7, pp 255–308
[22] Callaway E M 1998 *Annu. Rev. Neurosci.* **21(2)** 47–74
[23] Yabuta N H and Callaway E M 1998 *J. Neurosci.* **18(22)** 9489–9499
[24] Williams S R and Stuart G J 2002 *Science* **295** 1907–1910
[25] Song S, Sjöström P J, Reigl M, Nelson S and Chklovskii D B 2005 *PLoS Biol.* **3(3)** e68
[26] Lefort S, Tomm C, Floyd Sarria J C and Petersen C C 2009 *Neuron* **61(2)** 301–316
[27] Deutsch S and Deutsch A 1993 *Understanding the Nervous System* (New York, NY, USA: IEEE Press)
[28] Priesemann V, Wibral M, Valderrama M, Pröpper R, Le Van Quyen M, Geisel T, Triesch J, Nikolić D and Munk M H J 2014 *Front Syst Neurosci.* **8** 108
[29] Pruessner G 2012 *Self-Organised Criticality* (Cambridge, UK: Cambridge University Press)
[30] Griffiths R B 1969 *Phys. Rev. Lett.* **23(1)** 17–19
[31] Vojta T 2006 *J. Phys. A: Math. Gen.* **39** R143
[32] Muñoz M A, Dickman R, Vespignani A and Zapperi S 1999 *Phys. Rev. E* **59(5)** 6175–6179
[33] Levina A, Herrmann J M and Geisel T 2007 *Nat. Phys.* **3** 857–860
[34] de Andrade Costa A, Copelli M and Kinouchi O 2015 *J. Stat. Mech.*
[35] Kohn A 2007 *J. Neurophysiol.* **97(5)** 3155–3164
[36] Henkel M, Hinrichsen H and Lübeck S 2008 *Non-Equilibrium Phase Transitions* (Dordrecht, The Netherlands: Springer)

[37] Bonachela J A, de Franciscis S, Torres J J and Muñoz M A 2010 *J. Stat. Mech.* P02015

Journal of Physics: Conference Series **686** (2016) 012009 doi:10.1088/1742-6596/686/1/012009

Magnetic relaxation of nanoparticles with cubic and uniaxial anisotropies

Marcos J. Correia[1,2], Walther Schwarzacher[2], Edson Ferreira Chagas[2,3] and Wagner Figueiredo[1]

[1]Departamento de Física, Universidade Federal de Santa Catarina,88040-970, Florianópolis, Brasil.

[2]H. H. Wills Physics Laboratory, University of Bristol, BS8 1TL, Bristol, UK.

[3] Instituto de Física, Universidade Federal de Mato Grosso, 78060-900 - Cuiabá, MT,Brasil.

E-mail: wagner.figueiredo@ufsc.br

Abstract. We use Monte Carlo methods to simulate the influence of Brownian rotation on the magnetic properties of a system of single-domain magnetic nanoparticles with cubic and uniaxial magnetic anisotropies. The distinguishing feature of the system is a strongly temperature-dependent viscosity. Such a system has been realized experimentally using magnetic nanoparticles suspended in a freeze-concentrated cryoprotectant solution.

In this work we investigate magnetic phenomena related to the rotation of magnetic nanoparticles, by employing the Brown relaxation theory [1], during the transition between frozen and unfrozen states of the system. We developed a computational method to calculate the magnetic properties related to this process. We consider a system composed of single-domain magnetic nanoparticles dissolved in a solvent that also contains a cryoprotectant. On freezing the solvent, a thin liquid layer consisting of solvent and freeze-concentrated cryoprotectant forms around each nanoparticle. The glass transition temperature of this liquid layer T_g is lower than the freezing temperature of the solvent. In this way, while the solvent is already frozen, the freeze-concentrated cryoprotectant still permits nanoparticles to rotate. Initially, the system is found in a temperature below T_g, that is, without any rotational freedom. Increasing the temperature, the system passes through T_g, so that the nanoparticles become completely free to rotate. This behavior was observed experimentally in Reference [2], where magnetic measurements were used to monitor the behavior of single-domain magnetite nanoparticles doped with 2.5 % cobalt enclosed by the near-spherical cage protein ferritin (CMF) in a freeze-concentrated cryoprotectant solution.

In order to study this system through simulations we employed the Monte Carlo method along with the Metropolis algorithm. We calculate the energy of a new configuration of the system just after a small variation in the direction of the magnetic moment of a nanoparticle randomly chosen. Then, we compare this energy with the previous one to accept or not the new configuration of the system [3, 4, 5]. Besides this procedure we also included the change in energy due to the rotation of the nanoparticle in the viscous medium.

We consider a set of N non-interacting single-domain magnetic nanoparticles. We adopt the following effective energy model [3, 6, 7]

$$E_a = -\sum_{i=1}^{N} K_{ui}\,(\vec{S_i}.\vec{e_{1i}})^2 - \frac{1}{2}\sum_{i=1}^{N}\sum_{j=1}^{3} K_{ci}\,(\vec{S_i}.\vec{e_{ji}})^4 - \sum_{i=1}^{N} \vec{H}.\vec{\mu_i}, \tag{1}$$

where $K_{ui} = k_u V_i$, and $K_{ci} = k_c V_i$ are the uniaxial and cubic anisotropies of a particle of volume V_i, respectively. In this study the anisotropy density constants k_u and k_c are assumed to be both positive, and we also define the relative anisotropy parameter $k_{uc} = k_c/k_u$. The volume V_i of the ith particle is selected from a log-normal distribution. The magnetic moment of the ith particle is written as $\vec{\mu_i} = \mu_i \vec{S_i}$, where $\vec{S_i}$ is a unit vector, $\mid \vec{S_i} \mid = 1$, and $\mu_i = m_s V_i$, with m_s the particle magnetization. In the above equation the vectors $\vec{e_{ji}}$ $(j = 1, 2, 3)$ are unit vectors parallel to the cubic axes of the ith particle.

In the case of rotational Brownian motion we consider also the energy contribution E_r

$$\frac{dE_r}{dt} = \sum_{i=1}^{N} 2\pi r_i^3 \eta \omega_i^2, \tag{2}$$

where η is the viscosity, r is the hydrodynamic radius and ω is the angular velocity. For a liquid whose temperature is close to the glass transition T_g, the viscosity η can be described by the empiric equation of Vogel-Fulcher-Tammann (VFT) [8, 9, 10]

$$\eta = \eta_0 \exp\left[\frac{DT_0}{T - T_0}\right], \tag{3}$$

where T_0 corresponds to a temperature for which η is very high, η_0 is a reference viscosity, and D is a constant which depends on the structure of the system [11]. Therefore, in our calculations we need to take into account the total energy $E = E_a + E_r$, that is, the magnetic and rotational energy contributions.

Besides making a change in the direction of the magnetic moment of the nanoparticle, we also make a change in the direction of its magnetic easy axis. This is necessary in order to take into account the physical rotation of the nanoparticle in the viscous medium. After we change both directions, magnetic moment and easy axis, we compute the variation in the total energy ΔE. When $\Delta E < 0$, both changes are accepted. On the other hand, if $\Delta E > 0$, we generate a random number A, uniformly distributed between zero and one. If $A < \exp(-\Delta E/k_B T)$, the new configuration is accepted, if not, we keep the original configuration.

We consider a competition between the cubic and uniaxial anisotropies [15, 12, 13] as in the corresponding experimental CMF system. The size of the particles are selected from a log-normal distribution with mean value $V_0 = 1$ and standard deviation $\sigma = 0.25$, in agremeent with the experimental data [2, 12].

In Fig. 1 we give the values of the reduced magnetization, which is defined by $m = \mu H/K_0$, as a function of the reduced temperature $t = k_B T/K_0$, where $K_0 = k_u V_0$ and the viscosity is $\eta = 10^{-12}\exp[3/(t - 0.33)]$. The open squares represent the non-aligned system (cooled to $t = 0$ at zero field), while the full squares represent the aligned system (cooled at t $= 0$, but in a strong magnetic field, $h = 20$). For the aligned system we do not observe a blocking temperature, and this is due to the fact that all the magnetic moments were already aligned in the direction of the applied field. Increasing the temperature, we observe a decrease in the magnetization. For the non-aligned case, we see that the magnetization increases with the temperature up to the blocking temperature. When $t \geq t_b = 0.14$, both systems, aligned and non-aligned, show a decrease in the magnetization as a function of temperature. However, the magnetization of the aligned system is always higher than the non-aligned one up to the temperature $t = 0.45$. Close to $t = 0.45$, the magnetization of the aligned system drops abruptally, assuming the same value as the non-aligned one. At this temperature, which we denote by T_R, and which is three times larger than t_b, the value of the viscous energy E_r is close to the magnetic energy associated with the magnetic anisotropy allowing nanoparticles to rotate due to the rotational brownian motion. From this temperature on, both systems become magnetically equivalent, where all the easy magnetization axes point randomly in space.

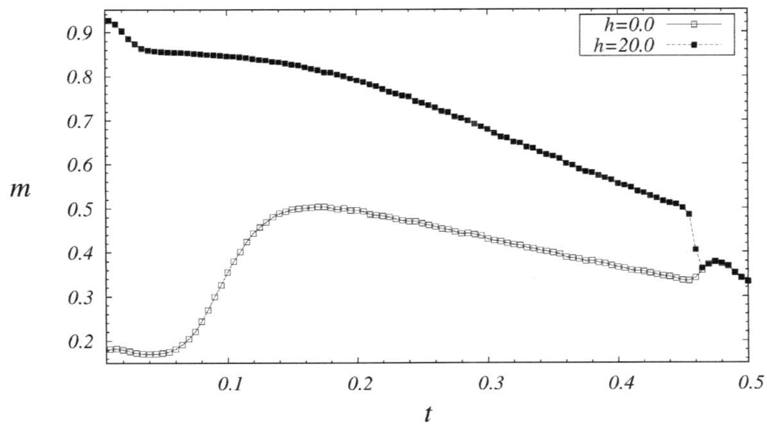

Figure 1. Reduced magnetization m as a function of temperature for a system with $\sigma = 0.25$, $k_{uc} = 2$, viscosity η. A small magnetic field $h = 0.5$ is applied during the heating process. The open squares represent the system initially cooled to $t \approx 0$ at zero field and the full squares represent the system initially cooled to $t \approx 0$ in an external field $h = 20$. The error bars are smaller than the size of the symbols, and the lines serve as a guide to the eyes.

Figure 2 shows the reduced magnetization m as a function of the external magnetic field h for two similar systems, but different cooling processes. In the first case, represented by the cross symbols, the system was cooled below T_0 ($t = 0.33$) in the presence of a strong magnetic field $h = 20$, from $t = 0.55$ to $t = 0.01$. In the second case the system was cooled below T_0 in a zero magnetic field, which is represented by the squares. We note a clear difference between the two hysteresis curves. The system cooled in a strong field keeps the easy magnetization axes partially aligned to the field, which results in larger coercive fields when compared to the system cooled in the absence of a field, in which the directions of the easy axes are completely randomly distributed.

The applied field during the cooling process causes the rotation of the nanoparticle, and this is the reason that we observe the alignment of the magnetization easy axes. The alignment is not complete because T_R is roughly three times larger than the blocking temperature of the system, as can be seen in Fig. 1. However, the alignment is sufficient to change the magnetic properties of the system. This feature is in agreement with the experimental result [2].

In Fig. 3, we show magnetization curves, when the system is cooled from high temperature, $t = 0.52 > T_R$ to $t = 0.42 < T_R$, in the presence of an external magnetic field h, then to t=0 in zero field. The field is turned off at $t = 0.42$, which is larger than t_b, so the magnetic moments can cross the energy barriers, and the system can relax to very small values of the magnetization. However, as the system is already immobile, their magnetization easy axes can not rotate leaving the system with a partial alignment dependent on the applied field h. We note in Fig. 3 that the blocking temperature does not depend on the system alignment, however,

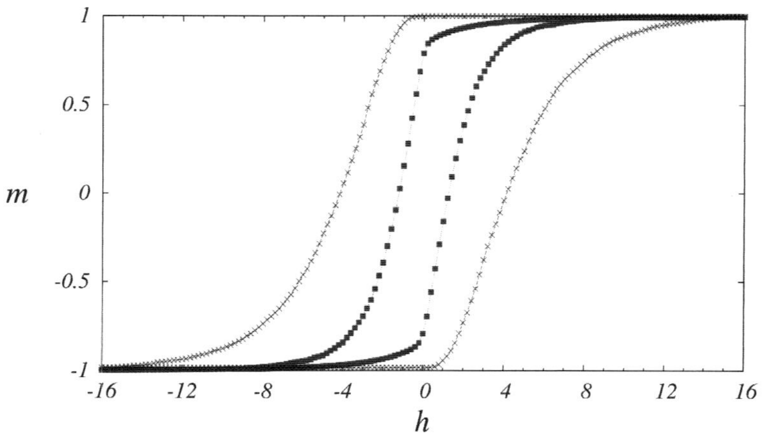

Figure 2. Hysteresis curves for the temperature $t = 0.02$ and $k_{uc} = 2$. The cross symbols, represent the system aligned by an external magnetic field, while the squares represent the system cooled in the absence of a field. The error bars are smaller than the size of the symbols, and the lines serve as a guide to the eyes.

the maximum value of the magnetization depends on the applied field during the cooling process, and after we reach the temperature T_R, all the curves have the same behavior as a function of temperature.

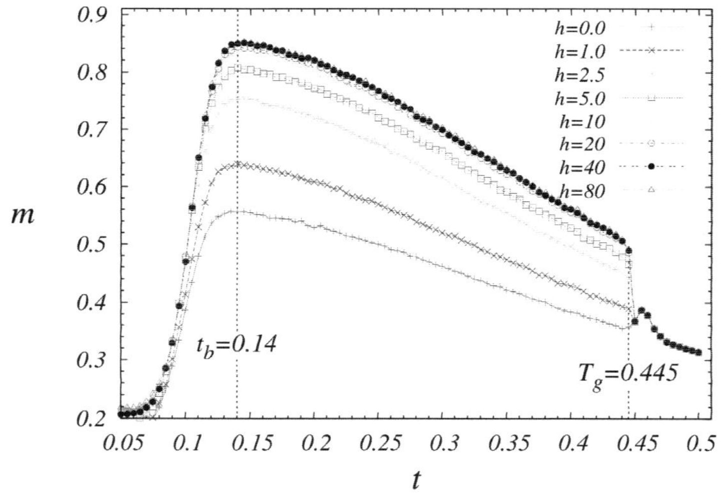

Figure 3. Magnetization curves for a system cooled in different values of the magnetic field (see text). We use the following set of parameters: $\sigma = 0.25$, $k_{uc} = 2$ and viscosity η. During the heating process we apply a small magnetic field $h = 0.5$. The error bars are smaller than the size of the symbols, and the lines serve as a guide to the eyes.

Figure 4 shows experimental magnetization curves for different values of the field applied during the cooling process. The system is similar to the one presented in Reference [2]. Qualitatively, the simulations shown in Fig. 3 and the experimental results agree very well.

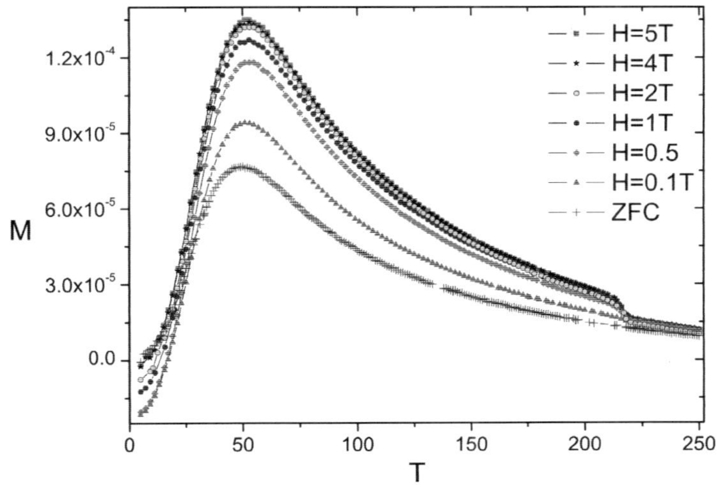

Figure 4. Experimental magnetization data for CMF cooled in the presence of different magnetic fields. Each system was cooled from temperature $250K$ to $200K$. During the heating process a small magnetic field $H = 0.05T$ was applied. Magnetic moments M are given in *emu* units, and temperature in K. The error bars are smaller than the size of the symbols.

The reduced magnetization as a function of the computational time (t_c), measured in Monte Carlo steps (MCs), for different temperatures, is shown in Fig. 5. In this case we have discarded the first 400 MCs, because in this time interval the Néel relaxation is dominant. The dashed black line is the fitting performed on the relaxation curves using the equation

$$y = M_0 \exp(-t_c/\tau_B), \tag{4}$$

where M_0 is a quantity related to the magnetization of the system after 400 MCs and τ_B is the Brown relaxation time. We consider the following temperatures: $t = 0.456$, $t = 0.458$, $t = 0.460$, $t = 0.462$, $t = 0.464$, and $t = 0.466$. From the fitting of these curves using eq. 4, we obtain the values of τ_B as a function of the temperature.

The behavior of the quantity $|\cos(\theta)|$, where θ is the mean angle made by the easy axis with the direction of the applied field, as a function of the number of Monte Carlo steps is shown in Fig. 6. At high temperatures and after a long time, the value of $|\cos(\theta)|$ reaches a minimum of around 0.55. This value is close to that expected for a completely random system, where $|\cos(\theta)| = 0.50$. In the figure the dashed line is the fitting employing the eq. $y = C_0 \exp(-t_c/\tau_B)$, where C_0 is related to the initial value of $|\cos(\theta)|$. We also found values of τ_B from the relaxation of $|\cos^2(\theta)|$, and the values of τ_B determined from this quantity, $|\cos(\theta)|$ and m are compared in Figure 7. In previous experimental studies [2] m was assumed to be proportional to $|\cos^2(\theta)|$ but from Fig. 7, though the values of τ_B determined from all the different quantities are similar, the values determined from m are closer to those determined from $|\cos(\theta)|$ than those determined from $|\cos^2(\theta)|$.

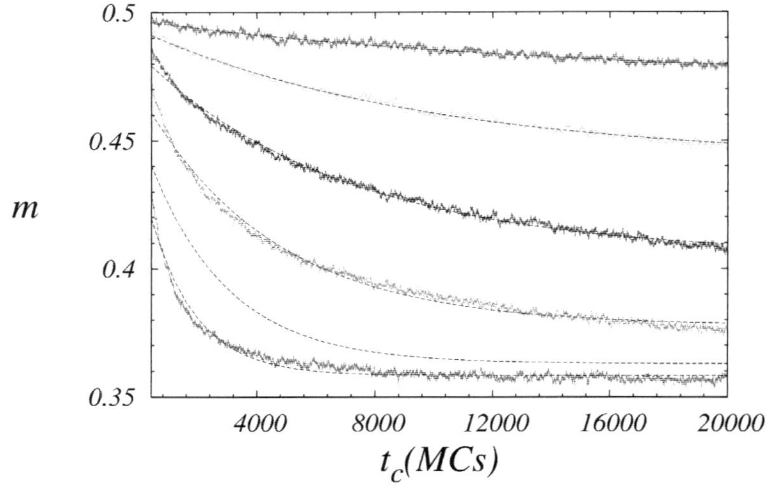

Figure 5. Reduced magnetization m as a function of the computational time measured in Monte Carlo steps in the presence of a small field $h = 0.5$. The parameters of the system are $\sigma = 0.25$, $k_{uc} = 2$. The system was cooled in the presence of a field $h = 20$. The temperatures are: $t = 0.456$, red curve, $t = 0.458$, green curve, $t = 0.460$, blue curve, $t = 0.462$, pink curve, $t = 0.464$, yellow curve and $t = 0.466$, grey curve. The dashed lines are fits using Eq. 4.

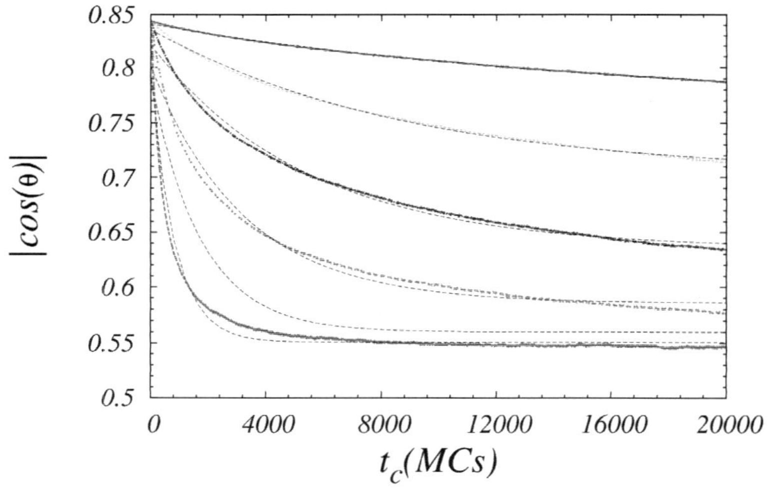

Figure 6. $|\cos(\theta)|$ (see text) as a function of the computational time measured in Monte Carlo steps in the presence of a small magnetic field $h = 0.5$, $\sigma = 0.25$, and $k_{uc} = 2$. The system was cooled in the presence of a field $h = 20$. Curves, red $t = 0.456$, green $t = 0.458$, blue $t = 0.460$, pink $t = 0.462$, yellow $t = 0.464$, and grey $t = 0.466$.

ACKNOWLEDGMENTS

M. J. Correia, Edson F. Chagas and W. Figueiredo would like to acknowledge the Brazilian agency CNPq for the financial support. This work was also partially supported by INCT-

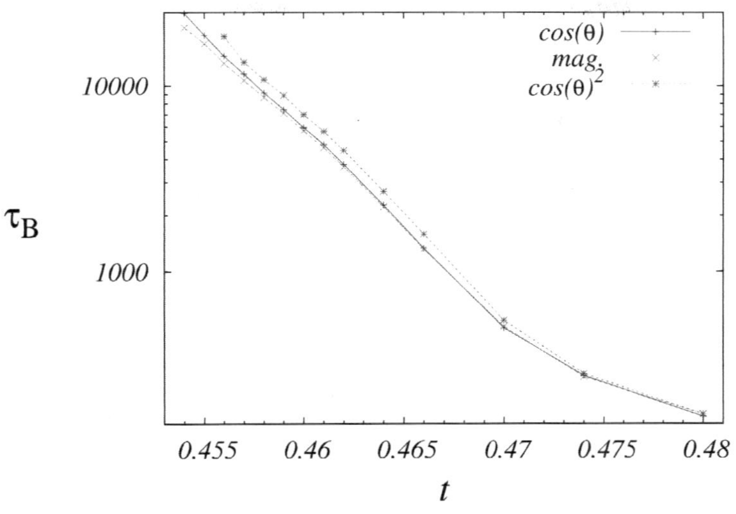

Figure 7. Determination of τ_B as a function of the reduced temperature, using three different procedures, as indicated in the figure.

FCX (FAPESP-CNPq). W. Schwarzacher would like to thank the Engineering and Physical Sciences research Council and the European Community's Seventh Framework Programme (FP7/2007–13) under Grant Agreement no. 228673(MAGNONICS).

References

[1] Brown W F 1959 *J. Appl. Phys.* **30** 130S
[2] Eloi J C, Okuda M, Jones S E and Schwarzacher W 2013 *Biophysical Journal* **104** 2681
[3] Figueiredo W and Schwarzacher W 2008 *Phys. Rev. B* **77** 104419
[4] Landau D P and Binder K 2009 *A Guide to Monte Carlo Simulations in Statistical Physics* (Cambridge University Press,Cambridge)
[5] Nowak U, Chantrell R W and Kennedy E C 2000 *Phys. Rev. Lett.* **84** 163
[6] Garanin D A and Kachkachi H 2003 *Phys. Rev. Lett.* **90** 065504
[7] Djardin P M, Kachkachi H, and Kalmykov Yu P 2008 *J. Phys. D: Appl. Phys.* **41** 134004
[8] Vogel H J 1921 *Physik Z* **22** 645
[9] Fulcher G S 1925 *J. Am. Ceram. Soc.* **8** 339
[10] Tammann G and Hesse W Z 1926 *Anorg. Allg. Chem.* **156** 245
[11] Dagdug L 2000 *J. of Phys. Cond. Matt.* **12** 9573
[12] Okuda M, Eloi J -C, Sarua A, Ward Jones S E, and Schwarzacher W 2012 *J. Appl. Phys.* **111** 07B519
[13] Eloi J -C, Okuda M, Carreira S, Schwarzacher W, Correia M J and Figueiredo W 2014 *J. Phys. Condens. Matter.* **26** 146006
[14] Néel L 1949 *Ann. Geophs. (C.N.R.S.)* **5** 99
[15] Correia M J, Figueiredo W and Schwarzacher W 2014 *Phys. Let. A* **378** 3366
[16] Hartmann U and Mende H H 1985 *Phil. M.* **52** 889-897

Nonequilibrium phase transition in directed small-world-Voronoi-Delaunay random lattices

F.W.S. Lima

Dietrich Stauffer Computational Physics Lab, Departamento de Física, Universidade Federal do Piauí, 64049-550, Teresina - PI, Brazil.

E-mail: `fwslima@gmail.com`

Abstract. On directed small-world-Voronoi-Delaunay random lattices in two dimensions with quenched connectivity disorder we study the critical properties of the dynamics evolution of public opinion in social influence networks using a simple spin-like model. The system is treated by applying Monte Carlo simulations. We show that directed links on these random lattices may lead to phase diagram with first- and second-order social phase transitions out of equilibrium.
Keywords: Opinion dynamics, Sociophysics, Majority vote, Nonequilibrium.

1. Introduction

Some collective or social behavior how workers decide whether or not to go on strike [1] or how a school of fish aligns into one direction for swimming [2] can be explained using the Ising model [3, 4]. The Ising model has been used to explain other behaviors as consensus formation [5], a fragmentation into many different opinions, or a leadership effect when a few people change the opinion of lots of others as also in the social impact model [6]. Other behaviors where opinions follow the majority of the neighbourhood and similar to it have been successfully explained using Ising-type model as the voter model of Liggett [7] and Schelling [5]. Most of these cited models and others can be found out in reference [8]. Beyond the Ising model many other models have been employed to study the social behavior of a set of individual or agents located on the nodes or sites of complex networks [9]. Some of these social systems also have been studied on complex networks of interacting agents. In most of these systems the interactions between the agents are directed, i.e., the links between agents act only in one direction (outwards or inwards). We investigate Sánchez-López-Rodríguez (SLR) [10] social model in two dimensions evolving on *directed* small-world-Voronoi-Delaunay (SWVD) random lattices in two dimensions with quenched connectivity disorder. The SRL model describes very well the dynamics of public opinion in social influence networks. In the present work we show that directed networks may lead to a highly nontrivial phase diagram including a social first- and second-order phase transitions out of equilibrium.

2. SLR Model on *directed* SWVD random lattice sizes

We consider a simple spin-like model (SLR), on *directed* SWVD random lattice by a set of spin variables S_i taking the values ± 1 situated on every site i of a *directed* SWVD random lattice with $N = L \times L$ sites, were L is the side of the square cluster. In this random lattice, similar to Sánchez et al. [10], we start from a two-dimensional SWVD random lattice consisting of sites linked to their k (where $3 < k < 20$ and different for each site of network) nearest neighbors by both outgoing and incoming links. Then, with probability p, we replace nearest neighbor outgoing links by new outgoing links to different sites chosen at random. After repeating this process for every link, we are left with a network with a density p of SWVD *directed* links. Therefore, with this procedure every site will have k outgoing links and varying (random) number of incoming links. Then, the spins or actors are then placed at the network sites. Then, any actor is connected by k outgoing links to other actors or mates and can be in one of two possible S_i states taking the values ± 1. Depending on the state of their mates and neighborhood, an actor may change its state according to a majority rule (ferromagnetic). In order to implement this, we introduce the payoff function:

$$G_i = 2S_i \sum_{j=1}^{k} S_j, \quad (1)$$

where the sum is carried out over the k mates of actor i. The external noise or social temperature T is included to allow some degree of randomness in the time evolution. Then, for a given value of the external temperature, the update of the model is then performed as follows: At each time step, an actor (network site) is randomly chosen and its corresponding G_i is computed according to Eq. (1).

- (1): If $G_i < 0$, $S_i \rightarrow -S_i$. The actor i is opposing the mates neighborhood majority and the change of its actual state is accepted.
- (2): If $G_i > 0$, the change of its current state is accepted with the probability

$$P_i = e^{-G_i/T}, \quad (2)$$

wich depends on temperature, i.e, an unfavorable change. Therefore, this model is a nonequilibrium one, since detailed balance is not satisfied.

The simulations have been performed on different *directed* SWVD random lattice sizes comprising a number $N = 250$, 500, 1000, 2000, 4000, 8000 and 16000 of sites. For each system size quenched averages over the connectivity disorder are approximated by averaging over $R = 100$ independent realizations. For each simulation we have started with a uniform configuration of spins. We ran 4×10^5 Monte Carlo steps (MCS) per spin with 2×10^5 configurations discarded to reach steady state. We do not see any significant change by increasing the numbers R and MCS. So, for the sake of saving computer time, the present values seem to give reasonable results for our simulation.

From the magnetization per spin, $m = \sum_i S_i/N$, we can derive the average the magnetization, the susceptibility, and the fourth-order magnetic cumulant,

$$M(T) = [< |m| >]_{\rm av}, \quad (3)$$

$$\chi(T) = \frac{N[< m^2 >_{\rm av} - < |m| >^2_{\rm av}]}{T}, \quad (4)$$

$$U_4(T) = 1 - [\frac{< m^4 >_{\rm av}}{3 < |m| >^2_{\rm av}}]. \quad (5)$$

In the above equations $< ... >$ stands for thermodynamic averages and $[...]_{\rm av}$ for averages over different realizations.

In order to verify the order of the transition of this model, we apply finite-size scaling (FSS). In the case of a first-order phase transition, we then expect, for large system sizes, an asymptotic FSS behavior of the form [11, 12],

$$\chi_{\rm max} = a_\chi + b_\chi N + c_\chi/N + ... \quad (6)$$

for the maximum of the susceptibility, $\chi_{\max}$, where a_χ, b_χ and c_χ are constants.

Otherwise, in the case of a second-order phase transition, we then expect, for large system sizes, an asymptotic FSS behavior of the form

$$[<|m|>]_{\rm av} = L^{-\beta/\nu} f_m(x)[1+...], \tag{7}$$

$$\chi = L^{\gamma/\nu} f_\chi(x)[1+...], \tag{8}$$

The $1/\nu$, β/ν, and γ/ν are the usual critical exponents ratio, and $f_i(x)$ are FSS functions with

$$x = (T - T_c)L^{1/\nu} \tag{9}$$

being the scaling variable. The dots in the brackets $[1+...]$ indicate corrections-to-scaling terms. We calculated the error bars from the fluctuations among the different realizations. Note that these errors contain both, the average thermodynamic error for a given realization and the theoretical variance for infinitely accurate thermodynamic averages which are caused by the variation of the quenched, random geometry of the lattices.

The correlation length exponent (ν) can be estimated from $T_c(L) = T_c + bL^{-1/\nu}$, where $T_c(L)$ is the pseudo-critical temperature for the lattice size L, T_c is the critical temperature in the thermodynamic limit, and b is a non-universal constant.

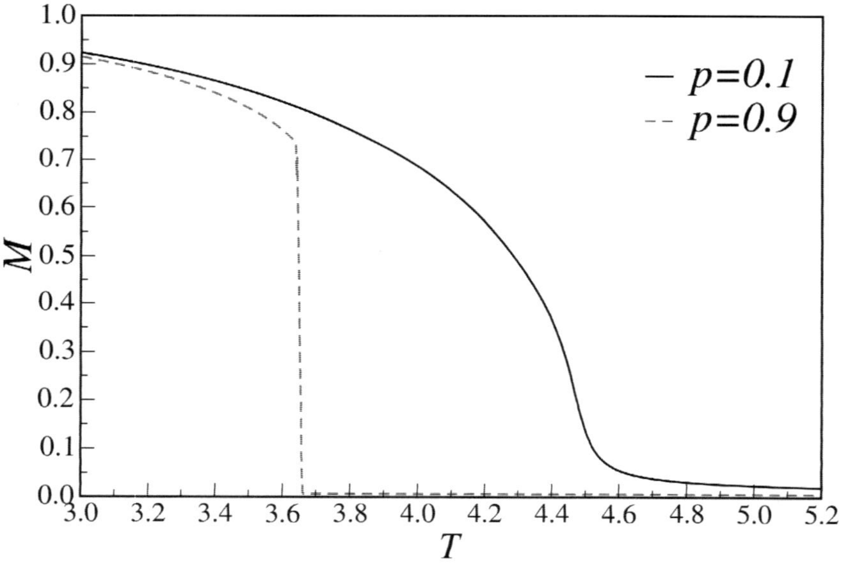

Figure 1. (color online) Magnetization as a function of T for $N = 16000$ and rewiring probabilities $p = 0.1$ and 0.9.

3. Results and Discussion

In Figure 1 we show the behavior of the magnetization versus temperature for $N = 16000$ and rewiring probabilities $p = 0.1$ and 0.9. One can see a typical behavior of a second and first-order phase transition for $p = 0.1$ and 0.9, respectively. In order to estimate the critical temperature we calculate the fourth-order Binder cumulant given by eq. (5). It is well known that these quantities are asymptotically independent of the system size and should intercept at the critical

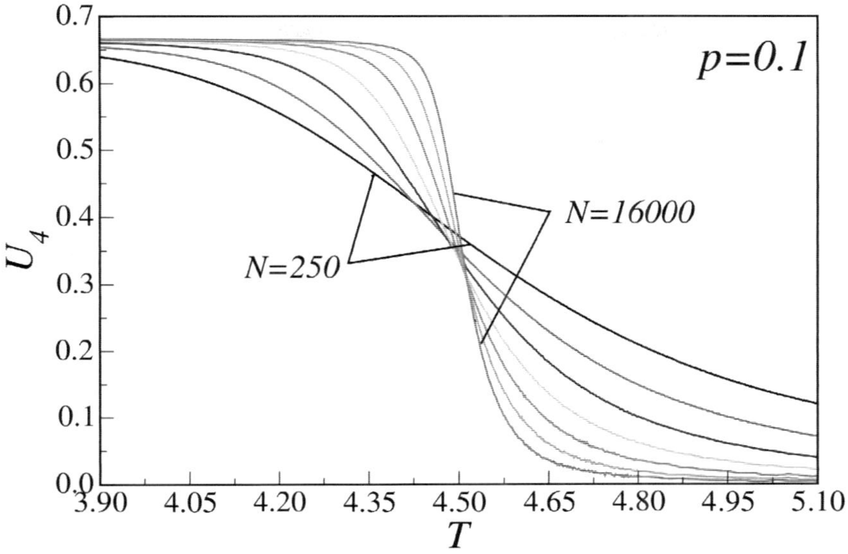

Figure 2. (color online) Plot of the fourth-order Binder cumulant as a function of T for various lattice sizes with $N = 250$, 500, 1000, 2000, 4000, 8000, and 16000 and rewiring probability $p = 0.1$.

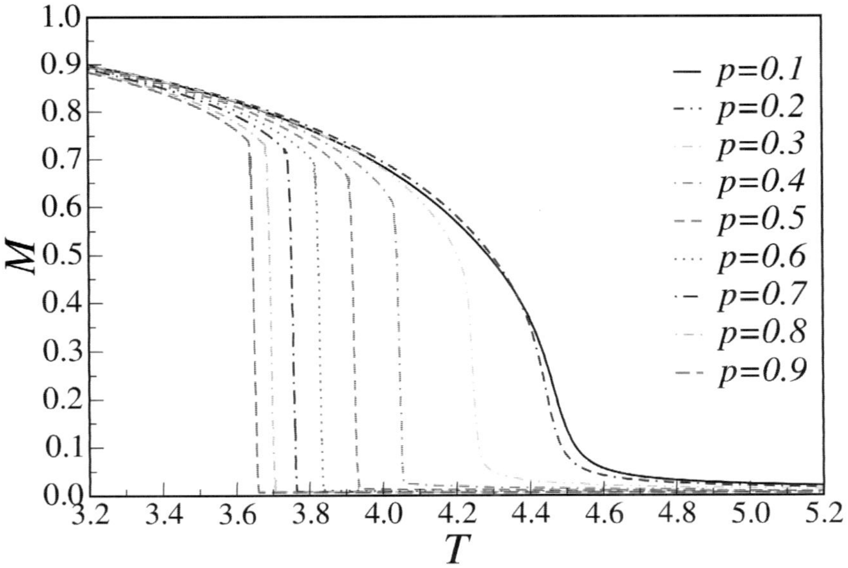

Figure 3. (color online) The same as Figure 1 for $N = 16000$ and various rewiring probabilities with $p = 0.1$, 0.2, 0.3, 0.4, 0.5, 0.6, 0.7, 0.8, and 0.9.

temperature [13]. In Figure 2 the fourth-order Binder cumulant is shown as a function of the T for several lattice sizes for the rewiring probability $p = 0.1$. Taking the largest lattices we have $T_c = 4.508(3)$ and $U_4^* = 0.342(3)$. One can see that U_4^* is different from the universal

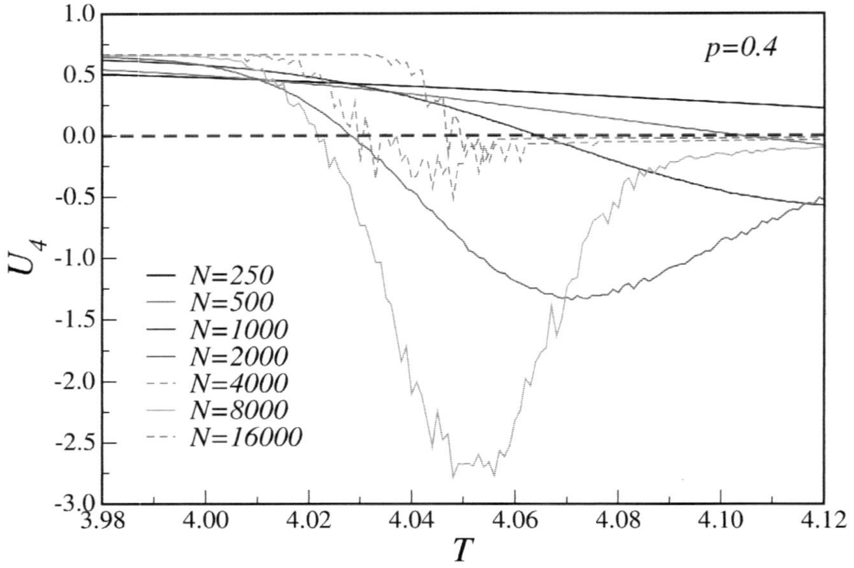

Figure 4. (color online) Same plot as Figure 2 for the fourth-order Binder cumulant as a function of T, but now for $p = 0.4$.

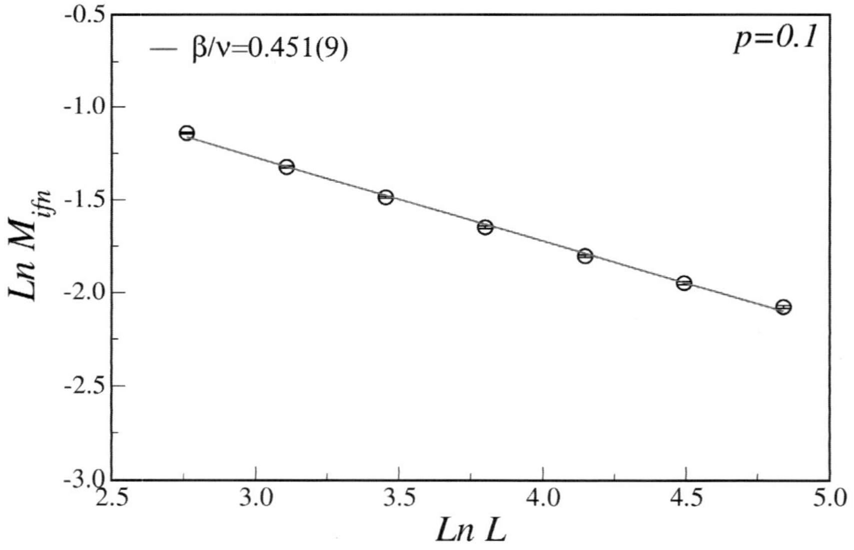

Figure 5. (color online) Plot of the logarithm of the modulus of the magnetization at the inflection point as a function of the logarithm of L. The solid line is the best linear fit.

value $U_4^* \sim 0.61$ valid for Ising models both on regular $d = 2$ lattices and on Voronoi-Delaunay random lattices [14, 15, 16].

In Figure 3 the magnetization as a function of T for $N = 16000$ and various rewiring probabilities with $p = 0.1, 0.2, 0.3, 0.4, 0.5, 0.6, 0.7, 0.8,$ and 0.9. From the above results

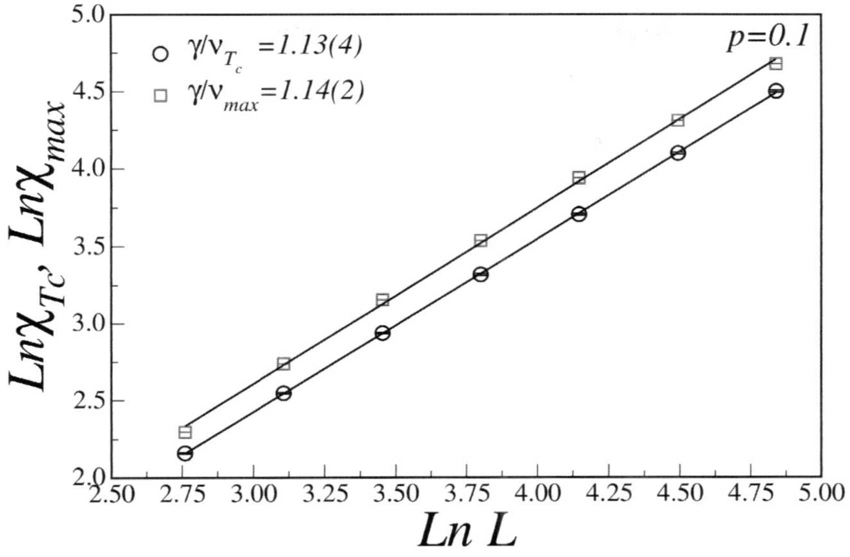

Figure 6. (color online) Log-log plot of the susceptibility χ at T_c and the susceptibility maxima $\chi_{\rm max}$ as a function of the logarithm of L. The solid lines are the best linear fits.

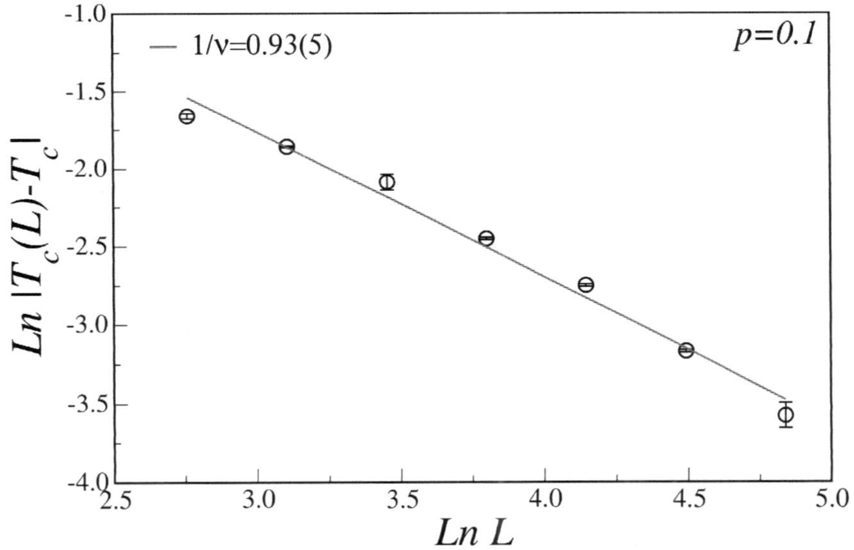

Figure 7. (color online) Plot the $\ln[T_c(L) - T_c]$ versus $\ln L$ for $p = 0.1$. The solid line is the best linear fit.

we can estimate the location of the tricritical point separating the first-order from the second-order transition. As for $p = 0.3$ one still has a second-order behavior and for $p = 0.4$ one has already a first-order transition, we could say that the tricritical point is located at $p = 0.35(5)$. Accordingly, the corresponding tricritical temperature is given by $T_t \approx 4.18$.

Figure 4 shows the fourth-order Binder cumulant as a function of T for $p = 0.4$ where one

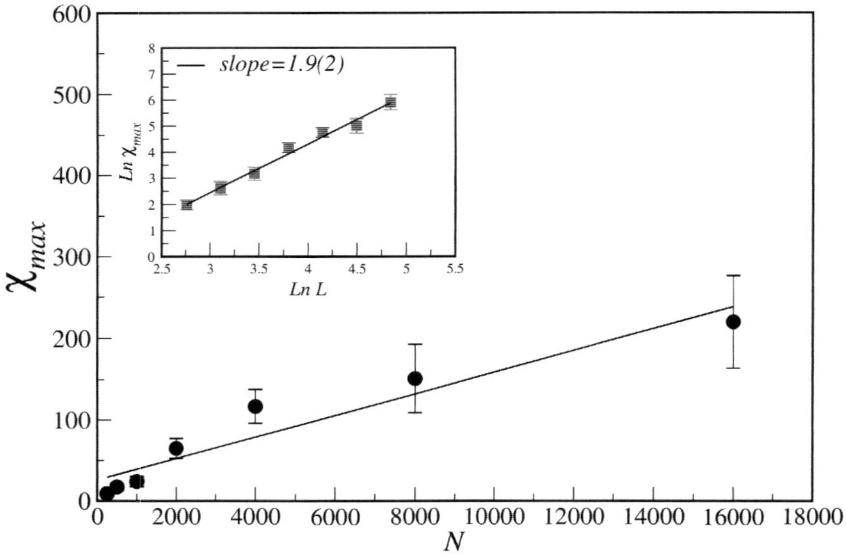

Figure 8. (color online) Plot of the susceptibility maxima $\chi_{\max}$ as a function of N for $p = 0.9$. The solid lines are the best linear fit. Note that in the inset we plot $Ln(\chi_{\max})$ versus $Ln(L)$, where $L = \sqrt{N}$.

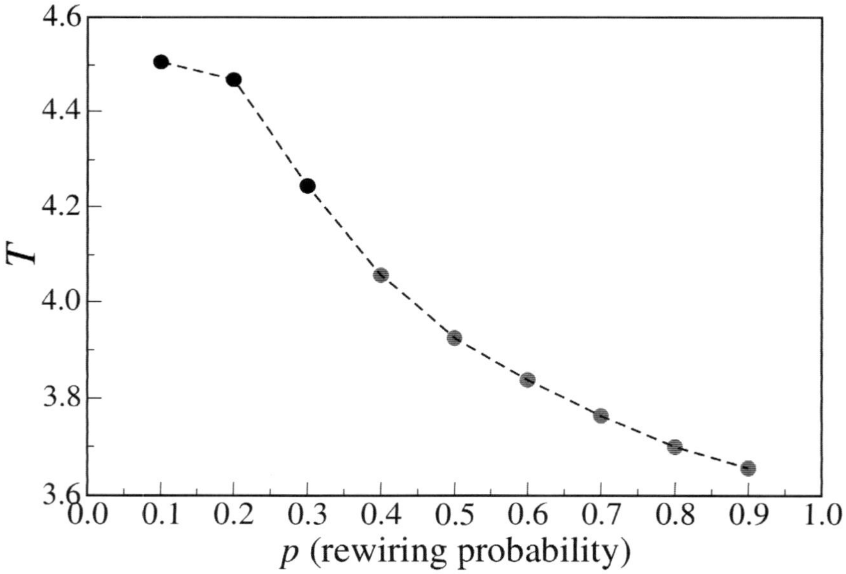

Figure 9. (color online) Plot of the transition temperature versus p for $p = 0.1$, 0.2, 0.3, 0.4, 0.5, 0.6, 0.7, 0.8 and 0.9. Note that for $p > 0.35$ the transition is of the first-order type.

has a first-order transition.

We have also computed the modulus of the magnetization at the inflection point and the magnetic susceptibility at T_c. The logarithm of these quantities as a function of the logarithm of L are presented in Figures 5 and 6, respectively. A linear fit of these data gives β/ν from the magnetization and γ/ν from the susceptibility. In addition, we plotted in Figure 6 the logarithm of the maximum value of the susceptibility $\chi_{\max}$ as a function of $\ln L$ for $p = 0.1$. One can also see that the exponents ratios $\beta/\nu = 0.451(9)$ and $\gamma/\nu = 1.13(4)$ are different from $\beta/\nu = 0.53(2)$ and $1/\nu = 1.06(5)$ of the SRL model and also of $\beta/\nu = 0.125$ and $\gamma/\nu = 1.75$ obtained for a regular $d = 2$ lattice, but they obey the hyper-scaling relation (within the error bars)

$$2\frac{\beta}{\nu} + \frac{\gamma}{\nu} = d, \tag{10}$$

where $d = 2$ for equilibrium models.

In Figure 7 it is shown a plot of $\ln[T_c(L) - T_c]$ as a function of $\ln L$ for $p = 0.1$. A linear fit of these data gives $1/\nu = 0.93(5)$. Figure 8 displays the maximum value of the susceptibility $\chi_{\max}$ as a function of N for $p = 0.9$. For a first-order phase transition we expect, for large system sizes, an asymptotic FSS behavior of the form given by eq. 6 [11, 12]. One can see the first-order nature of the transition. This fact is also illustrated in the inset of the figure where we have a log-log plot of the physical quantities as function of the lattice size.

Regarding the inset of Fig. 8, we have additionally done a linear fit using relations (6) with $c_\chi = 0$ and allowing the lattice dimension exponent vary as $N = L^d$ in those equations. We obtain, in this case, an exponent *slope* $= 1.9(2)$ for $p = 0.9$, which is indeed close to $d = 2$, as expected.

The phase diagram so obtained is depicted in Fig. 9 in the temperature T versus rewiring probability p. For $p \leq 0.35(5)$ we have a second-order phase transition, otherwise it is a first-order phase transition, indicating that there exists a nonequilibrium tricritical point at $p_c = 0.35(5)$.

4. Conclusion

In summary, the SRL social model on *directed* SWVD random lattice presents a second-order phase transition for $p \leq 0.35(5)$, otherwise it is a first-order phase transition with a nonequilibrium tricritical point at $p_c = 0.35(5)$. For $p = 0.1$ we find out the exponents ratios $\beta/\nu = 0.451(9)$, $\gamma/\nu = 1.13(4)$ and $1/\nu = 0.93(5)$ that are different for SRL social model on square lattice, where $\beta/\nu = 0.53(2)$ and $1/\nu = 1.06(5)$. Our results are also differents of Ising model 2d, where $\beta/\nu = 0.125$, $\gamma/\nu = 1.75$ and $\nu = 1$. Therefore, the SRL social model is not robust and the results depend on the topology of the system.

4.1. Acknowledgments

The author thanks D. Stauffer for many suggestions and fruitful discussions during the development of this work and also for reading this paper. We also acknowledge the Brazilian agency CNPQ-Brazil for its financial support. This work was also supported by the system SGI Altix 1350 the computational park CENAPAD.UNICAMP-USP, SP-BRAZIL.

[1] Galam S, Gefen Y and Shapir Y 1982 *J. Mathematical Sociology* **9** 1
[2] Callen E and Shapiro D 1974 *Physics Today* **27** 23
[3] Lenz W 1921 *Z. Phys.* **21** 613 , Ising E 1925 *Z. Phys.* **31** 235
[4] Baxter R J 1982 *Exactly solved models in statistical mechanics*, London, Academic Press
[5] Schelling T C 1971 *J. Mathematical Sociology* **1** 143
[6] Latané B 1981 *Am. Psychologist* **36** 343
[7] Liggett T M 1985 *Interacting Particles Systems*, Springer, New York
[8] Stauffer D, de Oliveira S. Moss, de Oliveira P M C and Sá Martins J S 2006 *Biology, Sociology, Geology by Computational Physicists* Elsevier Amsterdam ; Castellano C, Fortunato S and Loreto V 2009 *Rev. Mod. Phys.* **81** 591
[9] Watts D J and Strogatz S H 1998 *Nature* **393** 440
[10] Sánchez Alejandro D, Lopes Juan M , and Rodriguez Miguel A (2002) *Phys. Rev. Lett.* **88** 048701
[11] Janke W and Villanova R (1995) *Phys. Lett. A* **209** 179

Journal of Physics: Conference Series **686** (2016) 012010 doi:10.1088/1742-6596/686/1/012010

[12] Berche P E, Chatelain C, and Berche B (1998) *Phys. Rev. Lett.* **80** 297
[13] Binder K (1981) *Z. Phys. B* **43** 119
[14] Spriu D, Gross M, Rakow P E L, and Wheater J F (1986) *Nucl. Phys. B* **265** [FS15] 92
[15] Janke W, Katoot M, and Villanova R (1993) *Phys. Lett B* **315** 412
[16] Lima F W S, Moreira J E, Andrade Jr. J S, and Costa U M S (2000) *Phys. A* **283** 100 .

Journal of Physics: Conference Series **686** (2016) 012011 doi:10.1088/1742-6596/686/1/012011

Monte Carlo study of the Spin-1 Baxter-Wu model in a crystal field

M. L. M. Costa[1] **and J. A. Plascak**[2–4]

[1]Instituto de Tecnologia - ITEC, Universidade Federal do Pará, Campus Universitário do Guamá, Rua Augusto Corrêa, No. 1, 66075-110, Belém, PA, Brazil

[2]Departamento de Física, Centro de Ciências Exatas e da Natureza, Universidade Federal da Paraíba, Campus I, Cidade Universitária 58051-970, João Pessoa, PB, Brazil

[3]Departamento de Física, Universidade Federal de Minas Gerais, C. P. 702, 30123-970, Belo Horizonte, MG, Brazil

[4]Center for Simulational Physics, University of Georgia, 30602 Athens-GA, USA

E-mail: luciacosta@ufpa.br,pla@hal.physast.uga.edu

Abstract. The two-dimensional Baxter-Wu model with spin-1, in the presence of a crystal field, is studied by using Monte Carlo simulations. The standard single-spin-flip Metropolis algorithm is used to generate the configurations from which the order parameter, specific heat and magnetic susceptibility are measured. Single histogram techniques are used to get the results close to the phase transitions. The finite-size scaling procedure is employed in order to obtain the critical behavior. The simulations have shown that the critical exponents are different from the spin-1/2 case and are crystal field dependent.

1. Introduction

The Baxter-Wu model was first introduced by Wood and Griffiths [1] as a model which does not exhibit invariance by a global inversion of all spins. It is defined on a triangular lattice and can be described by the Hamiltonian

$$\mathcal{H} = -J \sum_{<ijk>} s_i s_j s_k, \tag{1}$$

where the exchange interaction J is positive and the sum is over all triangles made up of nearest-neighbor sites on the triangular lattice. For the spin-1/2 model, where $s_i = \pm 1$, the exact solution obtained by Baxter and Wu gives $k_B T_c/J = 2/\ln(1+\sqrt{2})$, $\alpha = \nu = \frac{2}{3}$ and $\gamma = \frac{7}{6}$ [2], where k_B is the Boltzmann constant, T_c is the critical temperature, and α, ν and γ are the critical exponents of the specific heat, correlation length, and susceptibility, respectively. The system has also been studied with quenched impurities by Monte Carlo [3] and Monte Carlo renormalization group approaches [4]. Conformal invariance studies [5, 6] have shown that the pure spin-1/2 Baxter-Wu and the four-state Potts models have the same operator content and are in the same universality class. More recently, the short-time critical dynamics has been investigated through the relaxation of the order parameter at the critical temperature by Monte Carlo simulations [7].

On the other hand, for spin values greater or equal to one there are neither exact solutions nor even much approximate results. It is the purpose of this work to study the Baxter-Wu model for the spin-1 case by using Monte Carlo simulations, where the variables s_i take now the values $s_i = -1, 0, 1$, and the Hamiltonian can be written as

$$\mathcal{H} = -J \sum_{<ijk>} s_i s_j s_k + \Delta \sum_i {s_i}^2, \tag{2}$$

where Δ is the crystal field interaction. This model has been treated by conformal invariance, finite-size scaling and mean-field renormalization group approaches, [8] and Monte Carlo simulations at zero crystal field [9]. It has been shown that the system belongs to a different universality class with exponents that depend on the value of Δ.

In this sense, we apply the histogram techniques together with the Metropolis simulation algorithm in order to investigate the thermal behavior of the spin-1 Baxter-Wu model defined by Eq. (2) by considering the specific heat, order parameter, and magnetic susceptibility. Our main interest is to obtain, through a finite-size scaling analysis, the phase transition temperature as well as the critical exponents of the model for some values of the reduced crystal field $D = \Delta/J$, namely $D = -1$ and $D = +1$.

2. Simulation background

The simulations have been carried out by employing the single-spin-flip Metropolis algorithm [10, 11]. In the course of the simulations we considered triangular lattices with linear dimensions $L \times L$ and fully periodic boundary conditions for system sizes of length $L = 21, 48, 60, 90, 108, 156, 210$. Due to the fact that the system has, in addition to the ferromagnetic phase (with all spins up), three different ferrimagnetic phases with three different sublattices (one sublattice with spins up, and spins down on the other two sublattices), the allowed values of L must be always a multiple of 3. In this way, all ground states of the infinite lattice would fit on any finite lattice. Following equilibration runs comprising up to 10^8 MCS (Monte Carlo steps per spin) were performed (each equilibration comprised 3×10^5 MCS). Histogram reweighting [12, 13] and finite-size scaling (FSS) techniques were used to precisely locate the second-order phase transition. Regarding the histograms, great care has been taken in order to assure the reliability of the extrapolated results for all lattice sizes.

The thermodynamic quantities that have been measured in our simulations are the order parameter, defined as the root mean square average of the magnetization of the three sublattices

$$m = \sqrt{\frac{{m_A}^2 + {m_B}^2 + {m_C}^2}{3}}, \tag{3}$$

where m_A, m_B, and m_C are the magnetizations of the different sublattices, the order parameter susceptibility defined as

$$\chi = \beta L^2 \left(\left\langle m^2 \right\rangle - \langle m \rangle^2 \right), \tag{4}$$

where $\beta = 1/k_B T$ (where $\langle ... \rangle$ means an average over the generated Monte Carlo configurations), and the specific heat

$$C = \beta^2 L^{-2} \left(\left\langle E^2 \right\rangle - \langle E \rangle^2 \right), \tag{5}$$

where $\langle E \rangle$ is the mean value of the energy.

In the case of a second-order phase transition, we then expect, for large system sizes, an asymptotic FSS behavior of the form

$$C = C_{reg}(T) + L^{\alpha/\nu} f_C(x) \left[1 + A_C(x) L^{-\omega} \right], \tag{6}$$

$$\chi = \chi_{reg}(T) + L^{\gamma/\nu} f_\chi(x) \left[1 + A_\chi(x) L^{-\omega} \right], \tag{7}$$

$$\frac{d\ln[\langle m^p \rangle]}{dT} = L^{1/\nu} f_p(x) \left[1 + A_p(x) L^{-\omega}\right], \tag{8}$$

where $C_{reg}(T)$ and $\chi_{reg}(T)$ are regular temperature dependent background terms, ν, α, and γ are the usual critical exponents, and $f_i(x)$, with $i = C,\ \chi,\ p$, are FSS functions with $x = (T - T_c)L^{1/\nu}$ being the scaling variable. The second term in the brackets approximate all the corrections to scaling by a single term, where ω is the leading correction-to-scaling exponent and $A_i(x)$ are non-universal functions (see, for instance, reference [14]). One should still emphasize that Eq. (6) is valid for $\alpha > 0$ only. Eq. (8) gives the maximum value of temperature derivative of logarithm of the mean value of the p power of the order parameter.

3. Results

In the limit $D = \Delta/J \to -\infty$ one recovers the spin-1/2 model and we have the results $k_BT_c/J = 2.269185...$, $1/\nu = 1.5$, $\alpha/\nu = 1$, and $\gamma/\nu = 1.75$. So, let us start discussing the case $D = -1$. Independent evaluation of the critical exponent ν, as obtained from Eq. (8), without any consideration of the critical temperature T_c, is shown in Fig. 1 for the maximum derivative of the logarithm of m and m^2 (although other powers of m can also be used).

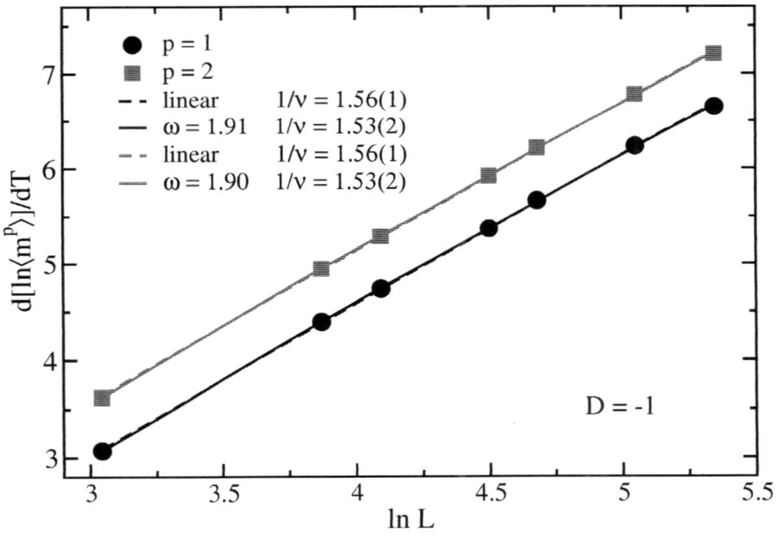

Figure 1. Logarithm of the maximum values of the derivatives of $\ln\langle m \rangle$ and $\ln\langle m^2 \rangle$ as a function of the logarithm of the size L for $D = -1$. The dashed lines are linear fits and the full lines are fits taken into account corrections to scaling. The errors are smaller than the symbol sizes.

The logarithms of the maximum values of the order parameter susceptibility and specific heat, as functions of the logarithm of L, are shown in Figs. 2 and 3, respectively.

From Figs. 1-3 one can see that for this value of the crystal field there is a slight deviation from the critical exponents of the spin-1/2 model. The results are still almost the same whether we take into account corrections to scaling or not. In most of the figures one cannot distinguish the linear fit from the one taking the corrections to scaling. We also note that there is not a strong dependence of the final fittings on the correction-to-scaling exponent, and in general the correction-to-scaling exponent remains close to the expected value of the spin-1/2 case, which is $\omega = 2$. The values of the critical exponents are, however, still closer to the expected results at the limit $D \to -\infty$.

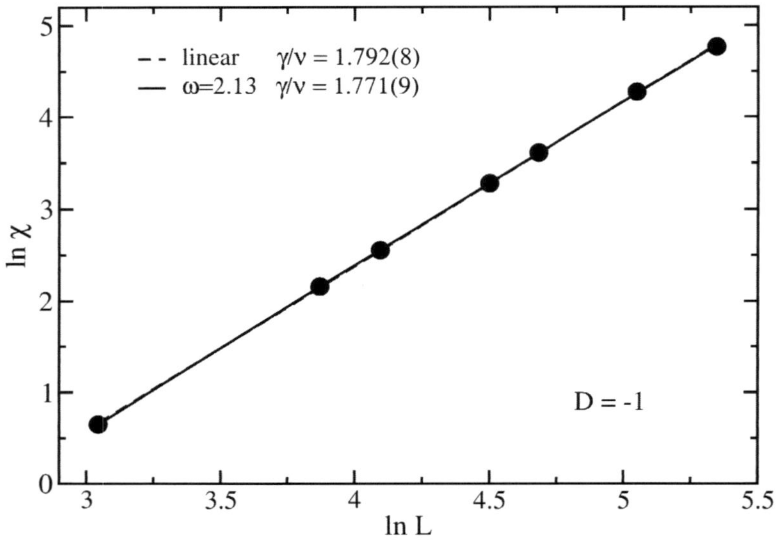

Figure 2. Logarithm of the maximum values of the order parameter susceptibility χ as a function of the logarithm of L for $D = -1$. The dashed line is a linear fit and the full line is a fit taken into account corrections to scaling. The errors are smaller than the symbol sizes.

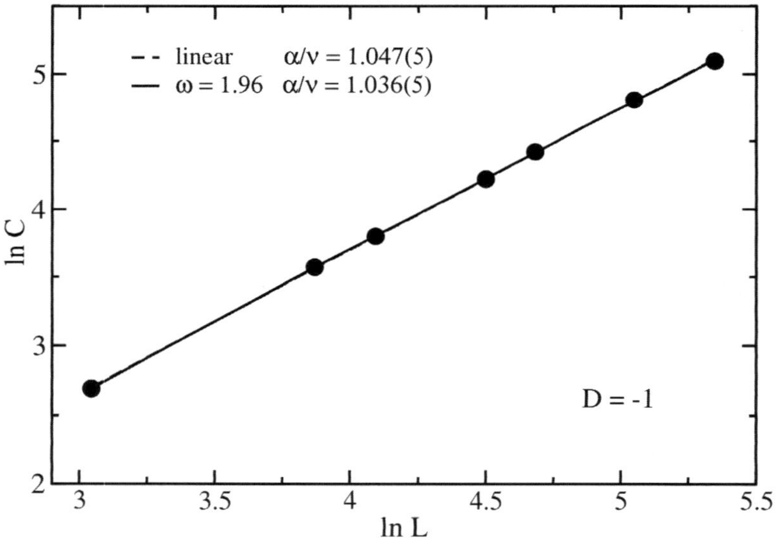

Figure 3. The same as Fig. 2 for the logarithm of the maximum values of the specific heat.

Having ν been determined quite accurately, we can proceed to estimate the position of T_c. As is well known, the location of the maxima of the various thermodynamic derivatives, namely the maximum of the specific heat, susceptibility, and the derivatives of $\ln \langle m \rangle$ and $\ln \langle m^2 \rangle$, provide estimates for the transition temperature itself. As the critical exponents have similar values by either considering or not corrections to scaling, in order to estimate the critical temperature we can use the simpler finite-size scaling relation

$$T_L = T_c + \lambda L^{-1/\nu} \;, \tag{9}$$

where λ is a constant, T_c is the critical temperature of the infinite system, and T_L is the effective transition temperature for the lattice of linear size L. A plot of these estimates is given in Fig. 4 for the four largest lattices we have studied and ν obtained from Fig. 1 by taking into account corrections to scaling. We can note that the critical temperature from the different quantities are indeed quite close to each other.

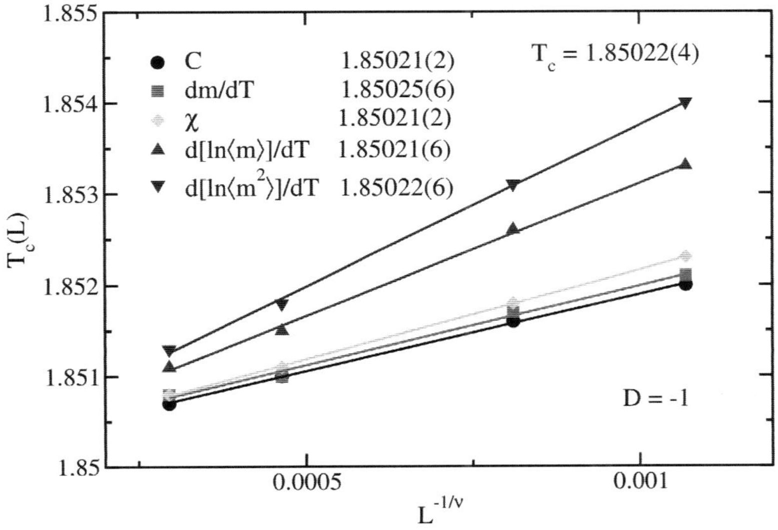

Figure 4. Size dependence of the effective critical temperatures (in units of J/k_B) estimated from several thermodynamic quantities. The lines are fits to Eq. (9) with ν obtained from Fig. 1 with corrections to scaling. The errors are smaller than the symbol sizes.

Now, Figs. 5, 6, and 7 show the corresponding results for the critical exponents for the case $D = +1$. One can see that the same behavior regarding the corrections to scaling is happening in this case, however, the critical exponents are now farther from the corresponding ones of the spin-1/2 model. Fig. 8 depicts the critical temperature obtained from several quantities for $D = +1$ and the four largest lattices considered.

Table 1. Estimated critical exponents and critical temperatures for different values of the crystal field. The results for $D = 0$ are from reference [9] and the exact ones for $D \to -\infty$ from reference [2]

D	$1/\nu$	γ/ν	α/ν	T_c
$-\infty$	1.5	1.75	1	2.269185..
-1	1.53(2)	1.771(9)	1.036(5)	1.85022(4)
0	1.621(5)	1.83(3)	1.12(1)	1.6607(3)
1	1.864(8)	1.951(5)	1.53(2)	1.35966(3)

4. Conclusions

It is clear, from the quality of the above results, that a well defined second order phase transition takes place in the model with critical exponents which are indeed different from the spin-1/2 case. This means that this three-spin interaction model has exponents which depend not only

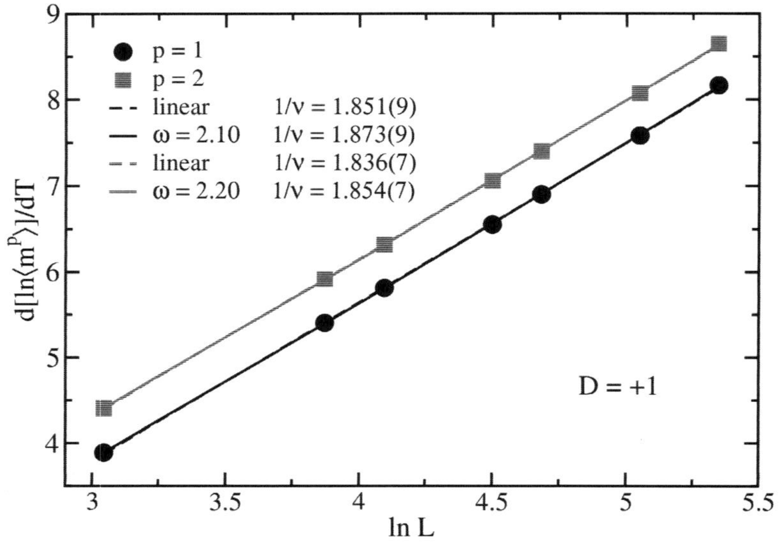

Figure 5. The same as Fig. 1 for $D = 1$.

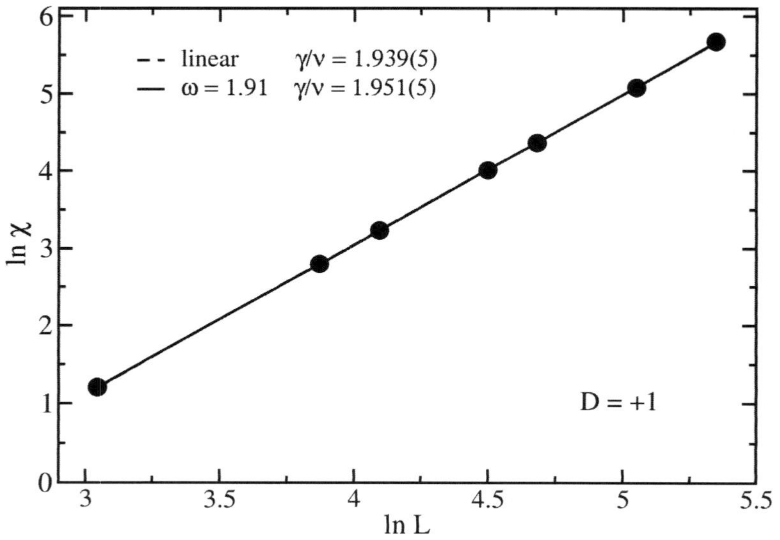

Figure 6. The same as Fig. 2 for $D = 1$.

on the spin value, but also on the crystal field. Table 1 shows the results for the present case in comparison to the values of the spin-1/2 model.

From the present results we then expect a second-order transition line whose critical temperature decreases as the crystal field increases, with varying critical exponents. This is indeed in contrast with the conjecture that the spin-1 Baxter-Wu model is critical only in the limit $D \rightarrow -\infty$ [15]. Moreover, the present results are in agreement with the picture of a line of second-order phase transition with varying critical exponents and the presence of a multicritical point, as has already been obtained from conformal invariance with finite-size scaling theory and the mean-field renormalization group approach [8]. Work in the direction of getting the

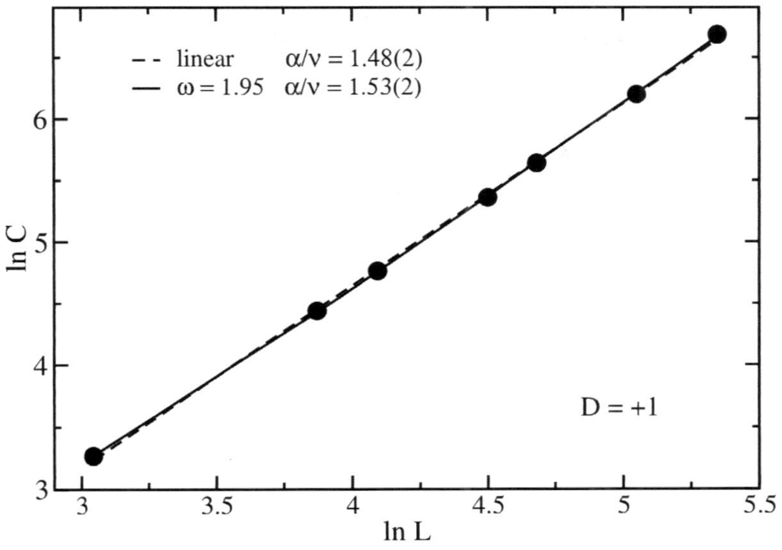

Figure 7. The same as Fig. 3 for $D = 1$.

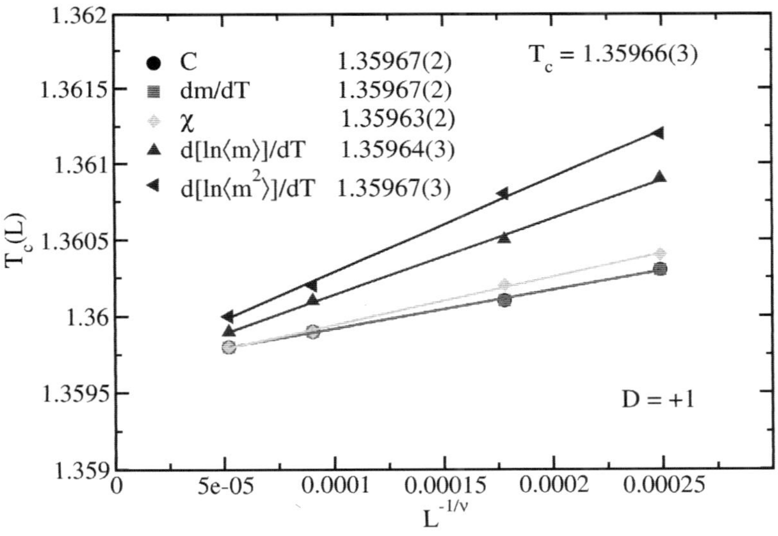

Figure 8. The same as Fig. 4 for $D = 1$.

multicritical point using Monte Carlo simulations is now in progress.

5. Acknowledgments
Financial support from CNPq and Capes are gratefully acknowledged.

References
[1] Wood D W and Griffiths H P 1972 *J. Phys. C* **5** L253
[2] Baxter R and Wu F 1973 *Phys. Rev. Lett.* **31** 1294
[3] Novotny M A and Landau D P 1985 *Phys. Rev. B* **32** 5874
[4] Novotny M A and Landau D P 1985 *Phys. Rev. B* **32** 3112

[5] Alcaraz F C and Xavier J C 1997 *J. Phys. A* **30**, L203
[6] Alcaraz F C and Xavier J C 1999 *J. Phys. A* **32** 2041
[7] Santos M and Figueiredo W 2001 *Phys. Rev. E* **63** 042101
[8] Costa M L M, Xavier J C and Plascak J A 2004 *Phys. Rev. B* **69** 104103
[9] Costa M L M and Plascak J A 2004 *Braz. J. Phys.* **34** 419
[10] Landau D P and Binder K 2014 *A Guide to Monte Carlo Simulation in Statistical Physics*, (4th edition, Cambridge University Press, Cambridge, England).
[11] Newman M E J and Barkema G T 1999 *Monte Carlo Methods in Statistical Physics*, Oxford University Press
[12] Ferrenberg A M and Swendsen R W 1988 *Phys. Rev. Lett.* **61** 2635; 1989 **63** 1195
[13] Ferrenberg A M 1991 *in Computer Simulation Studies in Condensed Matter Physics III*, edited by D. P. Landau, K. K. Mon, and H. -B. Schuttler (Springer-Verlag, Heidelberg)
[14] Landau D P 1994 Physica A **205** 41
[15] Kinzel W, Domany E, and Aharony A 1981 *J. Phys. A: Math. Gen.* **14** L417